NANOTECHNOLOGY SCIENCE AND TECHNOLOGY

NANOSTRUCTURED CONDUCTING POLYMERS AND THEIR NANOCOMPOSITES: CLASSIFICATION, PROPERTIES, FABRICATION AND APPLICATIONS

NANOTECHNOLOGY SCIENCE AND TECHNOLOGY

Additional books in this series can be found on Nova's website at:

https://www.novapublishers.com/catalog/index.php?cPath=23_29&seriesp=Nanotechnology+Science+and+Technology

Additional e-books in this series can be found on Nova's website at:

https://www.novapublishers.com/catalog/index.php?cPath=23_29&seriespe=Nanotechnology+Science+and+Technology

NANOTECHNOLOGY SCIENCE AND TECHNOLOGY

NANOSTRUCTURED CONDUCTING POLYMERS AND THEIR NANOCOMPOSITES: CLASSIFICATION, PROPERTIES, FABRICATION AND APPLICATIONS

UFANA RIAZ

AND

S.M. ASHRAF

Nova

Nova Science Publishers, Inc.

New York

For permission to use material from this book please contact us:
Telephone 631-231-7269; Fax 631-231-8175
Web Site: http://www.novapublishers.com

Library of Congress Cataloging-in-Publication Data
ISBN: 978-1-60876-943-8

Riaz, Ufana.
Nanostructured conducting polymers and their nanocomposites : classification, properties, fabrication, and applications / authors, Ufana Riaz and S.M. Ashraf.
p. cm.
Includes index.
ISBN 978-1-60876-943-8 (hardcover)
1. Polymeric composites. 2. Nanostructured materials. I. Ashraf, S. M. II. Title.
TA418.9.C6R52 2009
620.1'92--dc22

2009048915

Published by Nova Science Publishers, Inc. ✝ New York

CONTENTS

PREFACE

Dimensionality plays a critical role in determining the properties of matter.The nanostructure of a material is the key factor in the development of novel properties and in controlling the structure at the nano-level. Nanotechnology is therefore a highly promising field of the twenty-first century, which is expected to totally restructure the technological applications in the fields of semiconductors, inorganic as well as organic materials, energy storage, and biotechnology. The term "Nanotechnology" can be defined as the controlled manipulation of materials with at least one dimension less than 100 nm. This technology attempts to integrate chemistry, physics, materials science, and biology to create new material properties that can be exploited to develop facile processes for the production of electronic devices, biomedical products, high performance materials and consumer articles.. The traditional chemical industry has become a highly mature industry producing commodity products based on established technologies.However, to generate the market value of any new product requires emphasis on speciality materials possessing new functionalities obtained from benign processing technologies For the commercialization of nanotechnology based products, the production of unique performance and customer benefits should be at a reasonable cost. Hence the major challenges faced by the manufacturing industry include operations ranging from particle formation, processing, and fabrication to integration of the nanomaterial.The commercialization of nanotechnology is expected to boost wide technological development, improved quality of life and societal benefits around the world. The systematic approach of synthesis and device formation of nanotechnology based products comprises of the following major levels:

1. *Nano synthesis:* making nanoscale building blocks including nanoparticles, nanotubes, and nanostructures
2. *Nano characterization:* measuring and characterizing the basic properties of nanoscale building blocks, intermediate and final product forms.
3. *Nano processing and nanofabrication:* manipulating and processing nanoscale building blocks for a desired purpose and *incorporating* nanoscale building blocks , intermediate into final product forms such as polymer composites.

During last two decades, the field of organic conducting polymers (OCPs), which are also referred to as “synthetic organic metals” have undergone a dramatic transition from chemical curiosity to revolutionary new plastic materials which have attracted great attention of the researchers. There are a majority of reports dealing with OCPs involving the study of p-doped polymers: polyaniline (PANI), polypyrrole (PPy), poly(p-phenylene vinylene) (PPV), polythiophene (PTh) and their derivatives. The properties and applications of these extensively investigated systems relate to their semi-conducting nature, electro-activity, and plasticity. Specifically, nanoconducting polymers can be used to synthesize materials with a desirable combination of various electronic, optical and magnetic properties that are not achievable in other micro materials.They can be readily tailored at nanoscale with numerous functionalities exhibiting uniquely versatile properties. Furthermore, the ability to control the growth of the polymer chains in a living fashion allows chemists to fabricate monodisperse, structurally precise nano conducting polymers and their composites for further utilization. One of the most important characteristic of nano conducting polymers is their self assembled synthesis in a controlled manner, which helps to obtain materials with a nanostructure even at the molecular level. These materials are lately finding various applications in light emitting diodes (LED), electrochromic displays, smart windows, field effect transistors (FET), electromagnetic interference (EMI) shielding devices , energy storage products such as batteries, super capacitors and fuel cells. Nanoparticles of conducting polymers can be formed by chemical or electrochemical synthesis Traditional precipitation, crystallization, sol–gel transformation , emulsion polymerization are extended for nanoparticle synthesis. molecular template nucleation is used for specific size, morphology and control. They are generally produced in different forms such as wires, tubes, globules and discs. Conducting polymer nanowires (CPW) are attractive alternatives to silicon nanowires and carbon nanotubes because of their tunable conductivity, flexibility, chemical diversity, and ease of processing. They are prepared using a variety of protocols, such as chemical synthesis,templated electrochemical synthesis, and electro-spinning. The ability to build and control engineered polymeric materials at the nanometer (nm) scale is important for current and future development of materials for a wide range of applications. The phenomenal effect of these nanoparticles when combined with existing bulk material is evident in their becoming more lighter, stronger, and possessing higher ductility. The size of the nanoparticle grains strongly influences the properties in the bulk material. However, to overcome the poor chemical properties and processibility which limits their applicability nanoconducting polymers are mixed with other polymeric /non polymeric matrices to develop conducting polymer based nanocomposites.

Nanostructured conducting composites is a primer ingredient in microelectronics. It provides the possibility of further miniaturization as well as integration of devices at nano level which makes it more cost effective besides possessing the properties of high performance and lower power consumption. However, the assessment of the risk involved in using such nanomaterials represents certain ambiguity as scarce research is available based upon the beneficial conclusions and recommendations. A preliminary framework has been developed to help determine what research is needed, how it can be integrated, and how the resulting information can be incorporated into decisions about safety. The focus of this review is to highlight the state of nano conducting polymer and conducting polymer nanocomposites, the state of knowledge in processing, manufacturing, characterization, and the challenges in the integration of these nanomaterials for their potential applications in various fields.

Acknowledgement

The author Dr.Ufana Riaz also wishes to acknowledge the *Department of Science and technology (DST)-science and engineering research council (SERC),* India vide sanction no SR/FT/CS-012/2008 for granting the Young Scientist fellowship under the "Fast Track Scheme for Young Scientists".

Chapter 1

Introduction to Nanotechnology

The terms "Nanoscience and Nanotechnology" refer to the synthesis, characterization, exploration, and application of nanostructured materials. The prefix of "nanotechnology" derives from 'nanos' – the Greek word for dwarf, and is a billionth of a meter. The application of nanomaterials can be historically traced back to even before the generation of modern science and technology.Nanoparticles were anciently used as dye materials in ceramics [1]. Colloidal gold was used in medical treatment for the cure of dipsomania, arthritis etc, as early as in 19th century .Systemic experiments on nano-scale materials were initiated by Faraday in 1857 [2]. In 1959, Richard Feynman gave a lecture titled "*There's Plenty of Room at the Bottom*", suggesting the possibility of manipulating things at atomic level [3]. This was considered to be the dawn of nanotechnology. However, the real emergence of nanotechnology came in the early 1990s. Lately, the availability of sophisticated instruments for characterization of nonmaterials such as scanning electron microscopy (SEM), transmission electron microscopy (TEM) and scanning probe microscopy (SPM) have made the nano world more approachable to the researchers.

1.1. State-of-the-art of Nanotechnology

Nanotechnology has expanded extensively to other fields of interest due to the novel properties of nanomaterials discovered. Nanoparticles can be promisingly used in catalysts, functional coatings, nanoelectronics, nanophotonics, laser, solar cells, resonators and high sensitivity sensors, energy storage, drug delivery and biomedicines.

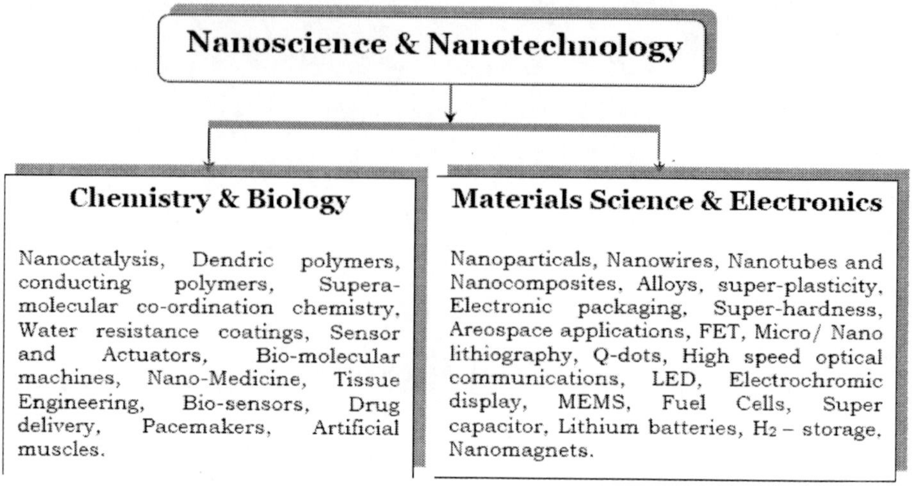

Figure 1. Representation of the interdisciplinary area of the science and technology of nanostructured materials and their applications.

Nanostructured thin films can be used in light emitting devices, displays and high efficiency photovoltaics.These are only limited areas of this rapid developing technology, as numerous other potential applications of nanomaterials have to be discovered. It has become an interdisciplinary research field in which many physicists, chemists, biologists, materials science and other specialists are involved. Nano investigations, being widely interdisciplinary by their very nature, promote the joining and merging of the various scientific and technological fields such as powder technology, colloid chemistry, surface chemistry and physics, clusters , and aerosols, tribology, catalysis, simulation and modeling, computer technique and so on .The science of nanotechnology has thus become an "umbrella" for a wide range of technologies and processes that can manipulate or exploit materials with an organized structure at the nanometer scale (1 nanometer is 10–9 meter).

1.2. Nanotechnology in the Chemical Industry – Opportunities and Challenges

We broadly see nanotechnology as an enabling technology, which promises to bring significant and sustainable advances to many market segments including chemical industries that require faster, smarter, and more convenient communication components and devices; smart materials that respond to external stimuli; cleaner energy, safer, efficient, and environmentally friendly manufacturing processes; improved human health diagnostic systems, and better ways to detect, prevent, as well as treat the detrimental consequences of chemical and biological agents. Commercialization of nanotechnology will lead to productivity improvements, improved health and improved quality of life resulting in societal benefits around the world. Nanotechnology is still very much in the melting-pot stage, but already developers are pursuing a range of opportunities that they believe will provide some key advantage over existing systems or processes as an intense mixture of creativity, innovation, and entrepreneurship in the world of nanotechnology.

Chapter 2

SYNTHESIS AND SELF-ASSEMBLY OF NANOMATERIALS

There are different types of nanostructured materials. These range from zero dimensional atomic clusters (quantum dots) to three dimensional structures, where at least one dimension is in the nanometer range, as shown in Figure 2.

Atoms, clusters, quantum dots are defined as "*Zero modulation dimensionality*" (0-D) systems which can have any aspect ratio ranging from 1 to ∞. Any wires, rods, belts and tubes in the nanometer range is classified as *One-dimensionally modulated* (1D) ,this way of classification provides a good system to investigate the dependence of electrical, thermal ,optical and mechanical properties on the dimensionality and size reduction (quantum confinement).Some of the established examples include size-dependent excitation or emission, quantized or ballistic conductance, coulomb blockade or single electron tunneling (SET) and metal-insulator transition. Many of these properties play an important role when interconnects and active components are fabricated in nanoscale electronic, optoelectronic, electrochemical, and electromechanical devices. The most important two-dimensional layers (2-D) in the nanometer thickness range consisting of ultrafine grains (nanometer range diameter) are "*Two dimensionally modulated*". This class includes insertion materials such as graphite ,layered-type dichalcogenides and oxides of the transition metals Ti, Nb, Ta, Mo, W and V, which are of interest as positive and negative insertion electrode materials for electrochemical power sources. Some of these materials also function as solid lubricants because of their layered character.

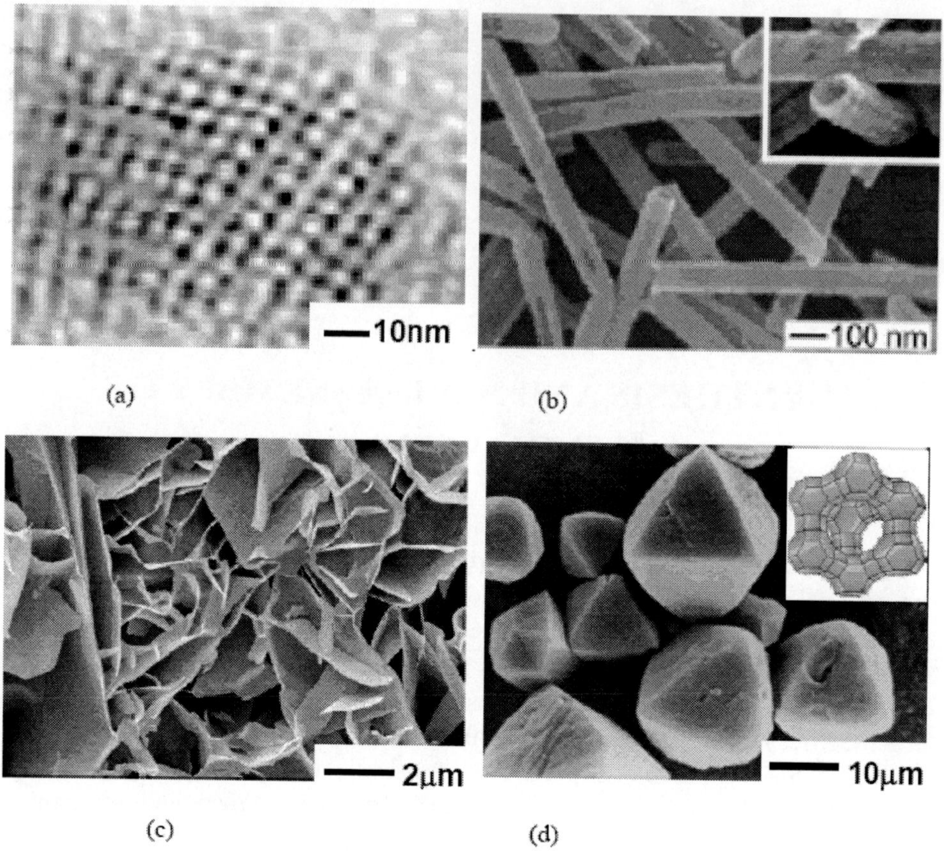

Figure 2. (a) 0-Dimensional quantum dots,(b) 1-Dimensional nanotbes (c) 2- Dimensional layered structure (d) 3-Dimensional Zeolites.

The last class consists of "*Three dimensionally modulated*" microstructures or nanophase framework where cross linked channels allow ion insertion if the size of the channels is sufficiently large to accommodate the ions. For example, 3-D-zeolitic solids containing aluminum, silicon, and oxygen with regularly spaced pores within the molecular framework can have nanopores enabling them to be useful for adsorbing small molecules, ions, or gases -just like how a sponge sucks up water. The pores are normally filled with positively charged ions, such as calcium or sodium which can be easily exchanged. Besides these types of individual nanostructures, ensembles of these nanostructures form high dimension arrays, assemblies, and superlattices. Nanomaterials have the structural features in between of those of atoms and the bulk materials. While most microstructured materials have similar properties to the corresponding bulk materials, the properties of nanomaterials are significantly different from those of atoms and bulks materials. This is mainly due to the nanometer size of the materials which enable them to have :

i. large fraction of surface atoms
ii. high surface energy
iii. spatial confinement
iv. reduced imperfections,

The above characteristics do not exist in the corresponding bulk materials [4]. Due to their small dimensions, nanomaterials have extremely large surface area to volume ratio, which makes a large fraction of atoms of the materials to be the surface or interfacial atoms, resulting in more "surface" dependent material properties.

Nanoparticles can be formed by either milling of large particles or by directed chemical synthesis.Small media mills and micro fluidizers are used to break down or disperse solids down to 100 nm.However most nanoparticles are directly synthesized from liquid or vapor phases. Chemical and physical vapor phase syntheses are well-established technologies for large-scale production of metal, metal oxides, and ceramic nanoparticles.Carbon black, color pigments, and fumed silica are perhaps the oldest commodity nanoparticles that have been widely used for centuries. High temperature flame, thermal plasma, or lasers are used as heat sources for vapor phase synthesis. Spray pyrolysis is a proven method to produce complex nanoparticles.Challenges for vapor phase synthesis encompasses the control of the particle size distribution (PSD), *in situ* formation of hybrid particles and structures, efficient collection systems, and continuous operations at high throughput and yield. Traditional precipitation, crystallization, sol–gel, emulsion polymerization are extended for nanoparticle synthesis. Molecular template nucleation is generally utilized for specific size, position, and crystal form control. Advances in these technologies have already proved to be practically important for dilute systems used in biocatalysis wherein filtration methods have been devised to deal with separations problems. Nanomaterials are, of course, abundant in nature as living organisms operate basically at a nanoscale level. Nanotechnologists seek to produce and utilize both novel nanomaterials and some natural nanomaterials in larger quantities and also within a more consistent size range.

Chapter 3

OVERVIEW OF INVESTMENT AND COMMERCIAL VALUE

Typically, much of the pre-commercial work in broad-use technologies has taken place in universities and academic institutions. Regional or federal funding is often the only source that can meet the significant costs of RandD projects at this stage. The United States in particular has made full use of its defense infrastructure to research and develop many technologies for both military and civilian applications. The value of nano-enabled products reached $166.6 billion in 2008; while global funding for nanotechnology mounted to over $25 billion of venture capital [5]. The European Union has also advanced research projects infrastructure to support burgeoning nanotechnology topics, as have other regions, including China, Russia, Switzerland, Australia, and Israel .Although both entrant and established companies are busy staking territory and promoting their particular view of nanotechnology, most advanced economies are funding efforts to establish regional capabilities during the next few years. Though much uncertainty currently exists, venture capitalists are treading their way into a complex area, and some are already making cautious investments—a trend that will likely continue in the next few years. Mainstream investors expect to hear increasingly about nanotechnology capabilities from both start-ups and established enterprises. Satisfyingly novel capabilities are also finding increased visibility on the platform of nanotechnology and—though perhaps years away from commercialization—could indeed eventually change lives. Nanotechnology has been weaving its unique threads into the technological tapestry. When the hoopla subsides, good science and good business opportunities will still be ahead.

Chapter 4

CONCEPT OF NANOCONDUCTING POLYMERS

An organic polymer that possesses the electrical, electronic, magnetic and optical properties of a metal/metal oxide while retaining the mechanical properties, processibility, etc. commonly associated with a conventional polymer, is termed an "intrinsically conducting polymer" (ICP) more commonly known as a "synthetic metal" ,Figure 3. In recent years, there has been agrowing interest in research on conducting polymer nanostructures (i.e., nano-rods, -tubes, -wires, and -fibers) since they combine the advantages of organic conductors and low-dimensional systems, and therefore create interesting physicochemical properties and promising useful applications [6–7]. After the exploration of conjugated polymers in the late 1970s, the chemical route of the science turned towards the nanosynthesis of these materials to enhance their intrinsic properties. Production of nano-shaped conducting polymers was first developed by Martin *et. al.*[8] using template synthesis method. One of the earliest applications of this method was in the preparation of nanoscopic electrodes within the pores of commercially available polymeric membranes [9].Both chemical and electrochemical synthetic methods have been used to the improve morphology of conjugated polymer nanotubes and nano-wires.Nano-synthesis of unidirectionally aligned polyaniline(PANI),polypyrole (PPy), poly(N-methylpyrrole) (PNMPy), polythiophene (PTh) and polyethylenedioxythiophene (PEDOT) in the form of nano-arrays, nano-tubes and nano-fibers using template synthesis method have been carried out [9-13] .PANI has always been used as a "*model material*" to systematically investigate the synthesis techniques , properties, and applications of nanofibers of conjugated polymers[14].

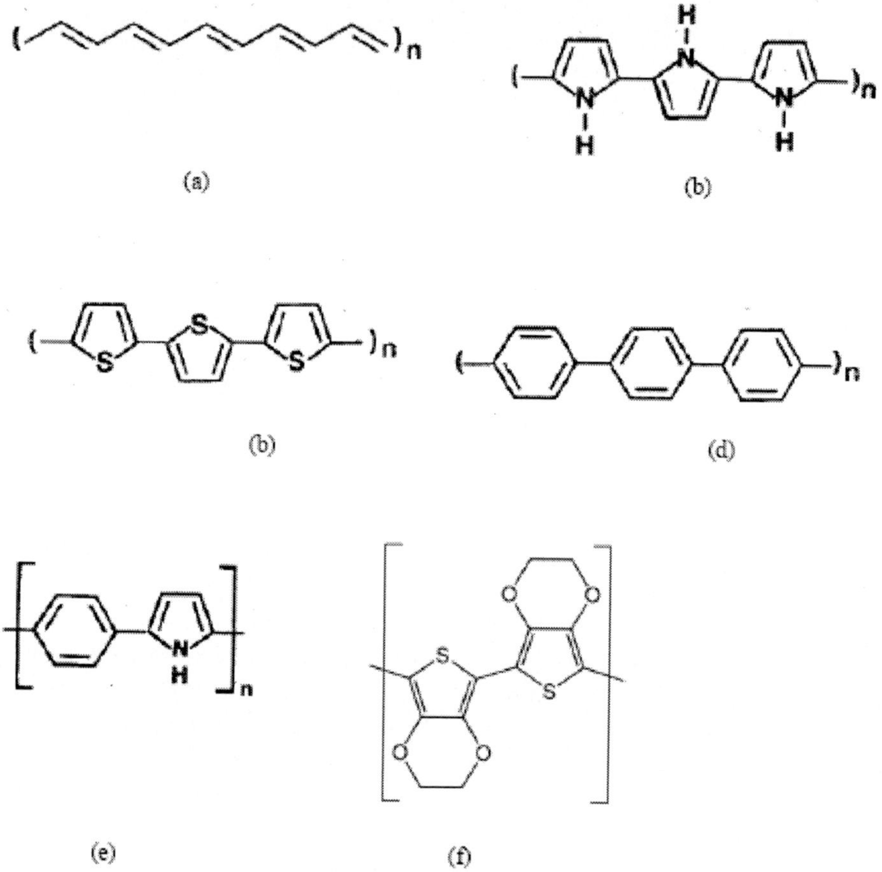

Figure 3. Structures of conducting polymers (a) polyacetylene (PA) (b) polypyrrole (Ppy) (c) polythiophene (PTh) (e) Polyaniline (PANI) (f) polyethylenedioxythiophene (PEDOT).

Innovative synthetic methodologies have been developed that readily produce high-quality, small-diameter nanofibers in large quantities. The impact of nanofibrillar morphology on the properties and applications of conducting polymers significantly improve their processibility and the performance which lead to much faster and more responsive chemical sensors, novel inorganic/polyaniline nanocomposites, and ultra-fast non-volatile memory devices. Additionally, highly conjugated polymeric structures produce new nanoscale phenomena that are not accessible with current organic systems. Thus, several methods have been developed to make nanostructures of conducting polymers that are discussed in the upcoming sections.

4.1. PREPARATION

4.1.1. Electrochemical Synthesis

The electrochemical method has been used for the synthesis of an array of nano-materials such as metals, inorganic composites, conducting polymer fibers or tubes and multi-metal striped rods [15-20]. The synthesis of conductive polymer nanotubes has been done chemically or electrochemically using various templates. Martin and co-workers pioneered the synthesis of conductive polymer nanotubes and nanofibers such as polypyrrole (PPy), polythiophene(PTh), and polyaniline (PANI) in the pores of a polycarbonate (PC) or alumina membrane(AO) [12,19-29]. The Wan group [28] introduced surfactants as a template as well as dopants to synthesize polyaniline and polypyrrole nanotubes.In this method, the conductive polymer is electrodeposited on the surface of electrospun nanofibers, which are removed to generate hollow conductive polymer nanotubes. Electrochemical alternating current (ac) method was designed for the synthesis of polypyrrole (PPy) nano-tubule arrays, Figure 4(a)and (b).

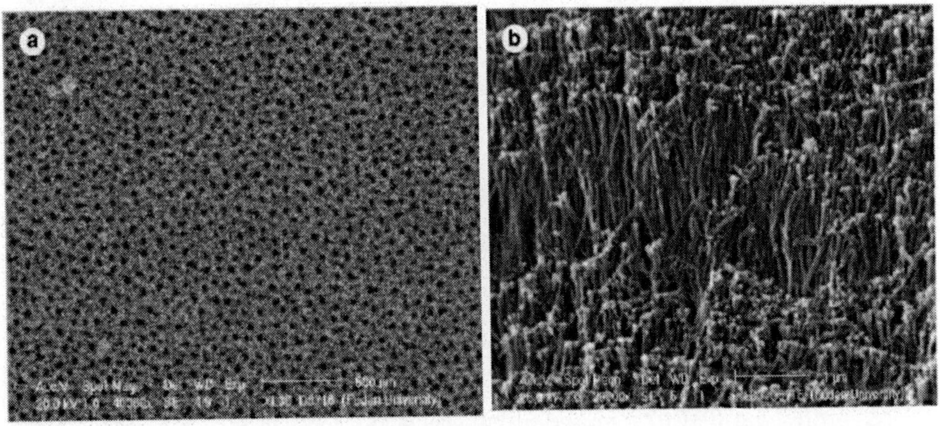

Reproduced with permission from Elsevier "Characteristics of polypyrrole (PPy) nano-tubules made by templated ac electropolymerization " L.Liu, C. Zhao, Y. Zhao, N.Jia, Q.Zhou, M.Yan and Z.Jiang, Eur.Polym.J.41,2005, 2117-2121

Figure 4. SEM images of the top of AAO/PPy membrane (a) fabricated by electrochemical ac method (polymerization voltage: 10 V, time: 8 h) and PPy nano-tubules (b) after the partial dissolution of alumina in 0.2M CrO_3 + 0.5 M H_3PO_4 solution.

The growth of nanotubes on a template can be explained by the mechanism based on the interaction, such as solvophobic and electrostatic, between the conductive polymer and the template [30]. Conducting polymer nanofibrils of ~100 nm in diameter can form "naturally" during electrochemical polymerization on the surface of the electrodes [31-32]. Studies indicate that uniform nanofibers can be obtained without the need for any template simply by controlling the electrochemical polymerization kinetics [33-37].However; the means of controlling their morphology without the use of templates are so far rather limited. Liu and co-workers [38] devised a three-step electrochemical route for the synthesis of polyaniline nanowires using polyacrylic acid as template. This polymer could be fabricated both by chemical and electro-chemical polymerization (ECP) in several solvents like acetonitrile or water.

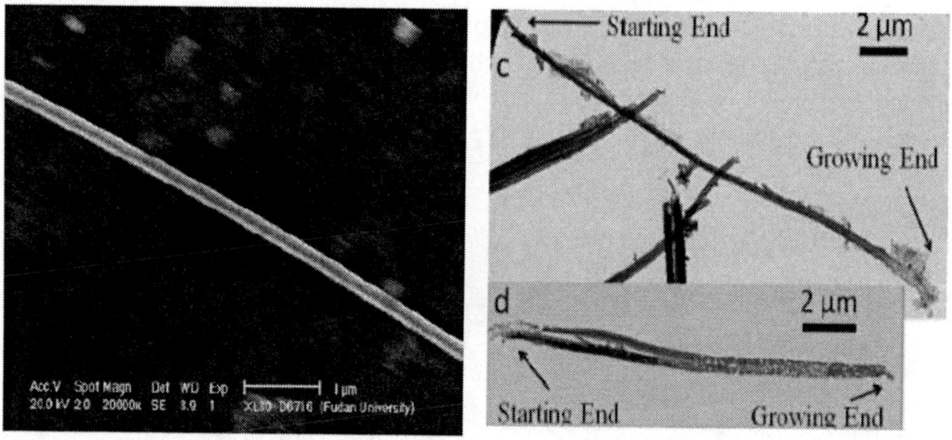

(a) (b)

(a). Reproduced with permission from Elsevier "Characteristics of polypyrrole (PPy) nano-tubules made by templated ac electropolymerization " L.Liu, C. Zhao, Y. Zhao, N.Jia, Q.Zhou, M.Yan and Z.Jiang, Eur. Polym. J., 2005, 41,2117-2121.

(b). Reproduced with permission from ACS "Morphology of Template-Grown Polyaniline Nanowires and Its Effect on the Electrochemical Capacitance of Nanowire Arrays" Y.Cao and T.E. Mallouk, Chem. Mater. 2008, 20, 5260–5265.

Figure 5. (a) SEM image of single PPy nano-tubule fabricated by ac method, polymerization voltage: 10 V with frequency of 50 Hz, 35 min in 20 wt% sulphuric acid solution.(b) TEM images of polyaniline nanowires grown in 0.25 and 1.5 M sulphuric acid solutions 900 mV.

The conducting polymer nanowires transfer electrical signals from one surface of the isolating support layer to the other surface. By using an ECP process, the growth of the polymer takes place within the membrane pores attached with the homogeneous electrode. In order to control the polymer growth, a modified ECP allows an external control of the polymer growth to avoid any wire interconnections over the surface and to have a maximal signal transport through the membrane, Figure 5(a) and (b) .By using a structured cathode, which consists of conducting metal pads surrounded by isolating areas, a controlled local growth of the polymer starting from the metallic pads on the cathode can be established. The synthesized nanotubes, nanofibers, and nanowires made by these methods are either interconnected or non uniform. The galvanostatic charge-discharge method is used to measure the electrochemical capacitance of arrays of nanowires, correlates well with their morphology. Transient current data for nanowire growth can be interpreted quantitatively in terms of instantaneous and progressive nucleation, diffusion and growth models, Figure 6, and this analysis is consistent with the observed concentration, and potential dependence of nanowire versus nanotube growth, Figure 7.

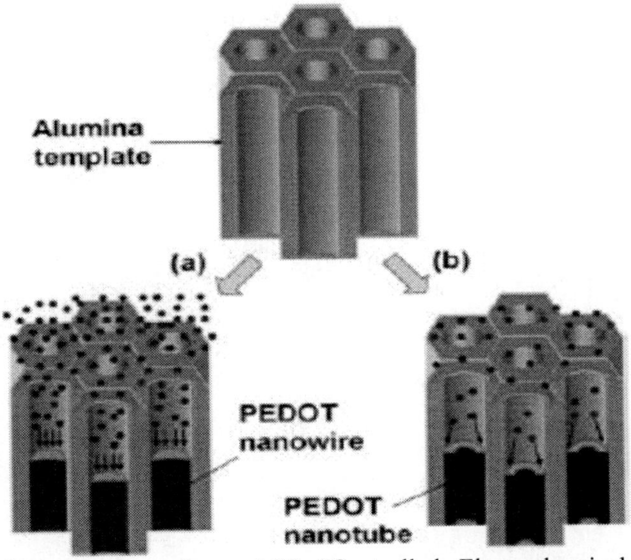

Reproduced with permission from ACS "Controlled Electrochemical Synthesis of Conductive Polymer Nanotube Structures" R.Xiao, S.Cho, R.Liu, and S.B.Lee, J. Am.Chem.Sco. 2007, 129, 4483-4489.

Figure 6. Growth Mechanism of PEDOT Nanostructures Based on Diffusion and Reaction Kinetics.

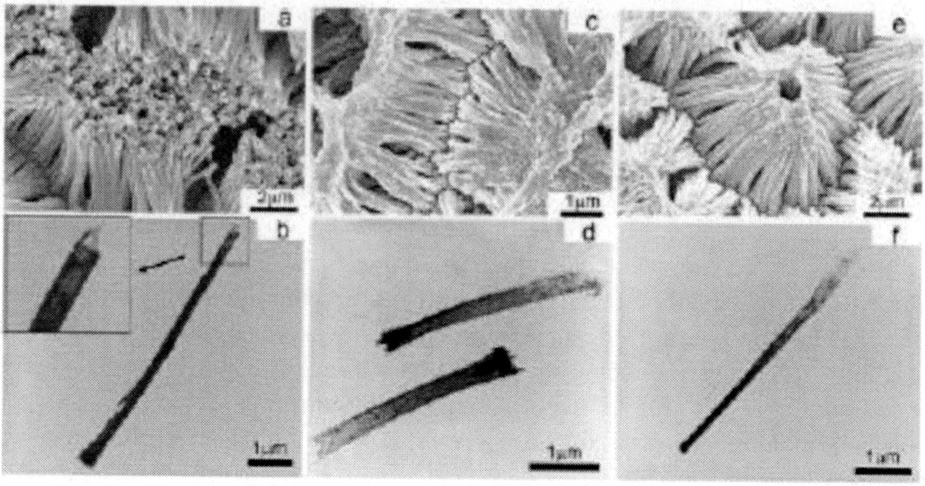

Reproduced with permission from ACS "Controlled Electrochemical Synthesis of Conductive Polymer Nanotube Structures" R.Xiao, S.Cho, R.Liu, and S.B.Lee, J.Am.Chem.Soc., 2007, 129, 4483-4489.

Figure 7. SEM and TEM images of PEDOT nanostructures synthesized potentiostatically (a,b) in 50 mM EDOT at 1.6 V, (c,d) in 50 mM EDOT at 1.2 V, and (e,f) in 25 mM EDOT at 1.5 V for 100 s. The upper (a, c, and e) and lower (b, d, and f) images were taken by SEM and TEM, respectively.

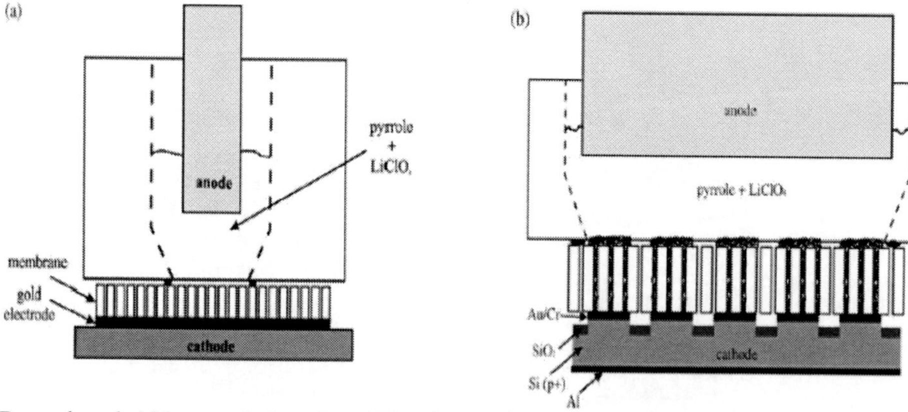

Reproduced with permission from Elsevier "Micro-patterning of self-supporting layers with conducting polymer wires for 3D-chip interconnection applications" J. Ackermann, C. Videlot, T.N. Nguyen, L. Wang, P.M. Sarro,D. Crawley, K. Nikolic, M. Forshaw, Applied Surface Science 212–213 (2003) 411–416.

Figure 8. Direct electrochemical polymerization of polypyrrole.

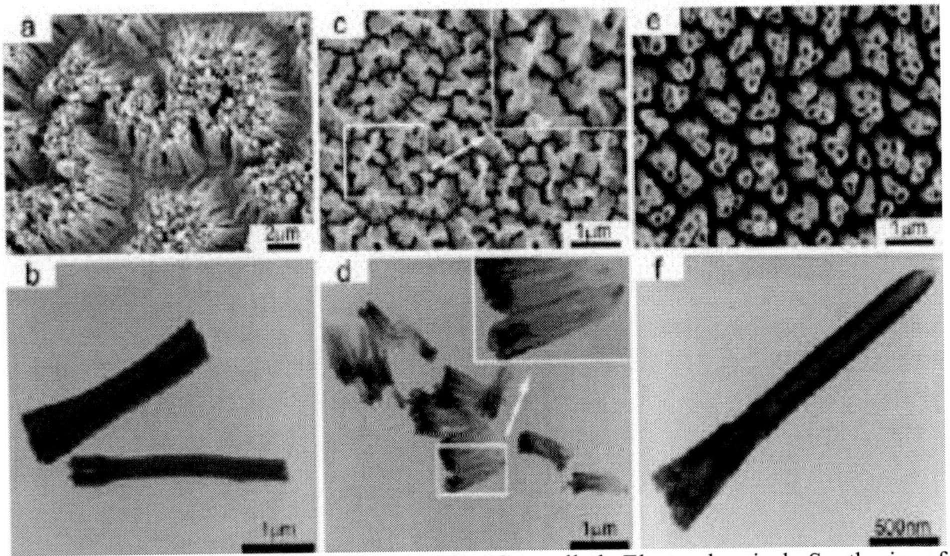

Reproduced with permission from ACS "Controlled Electrochemical Synthesis of Conductive Polymer Nanotube Structures" R.Xiao, S.Cho, R.Liu, and S.B.Lee, J.Am.Chem.Soc., 2007, 129, 4483-4489.

Figure 9. SEM of polypyrrole nanostructures synthesized potentiostatically for 100 s (a,b) in 50 mM pyrrole at 1.4 V, (c,d) in 10 mM pyrrole at 0.7 V, and (e,f) in 25 mM pyrrole at 1.8 V. The upper (a,c, and e) and lower (b, d, and f) images were taken by SEM and TEM, respectively.

The directional electrochemical polymerization (DEP) technique is a powerful method to pattern the self-supporting layers, Figure 8.In this process anodic aluminum oxide (AAO) is used with the ordered straight nano-channel the length and the pore diameter of the AAO membrane can be adjusted by controlling the anodizing voltage, temperature and selecting electrolytes such as oxalic, sulfuric, and phosphoric acid solutions, Figure 9 [39-40].

4.1.2. Chemical Synthesis

The conventional chemical oxidative polymerization of aniline is carried out in an aqueous solution in which aniline is dissolved in a strong acidic solution (for example, 1m HCl) at about 0^0C and the polymerization is initiated by adding an oxidant (for example, ammonium peroxydisulfate) into the solution. Careful analysis of polyaniline made in a conventional single-phase polymerization reaction by microscopy have shown the presence of a small amount of nanofibers

[41] .This observation indicates that the formation of nanofibers is not necessarily determined by the interface but more likely by the nature of the oxidative polymerization of aniline Polyaniline nanofibers form at an early stage in the polymerization process, Figure 10. These nanofibers have average diameters of 30–35 nm, Figure 10, which is consistent with those obtained using interfacial polymerization. As more ammonium peroxydisulfate is fed into the reaction, the nanofibers become scaffolds for secondary growth of polyaniline Figure 11 and finally turn into irregularly shaped agglomerates containing nanofibers and particulates.

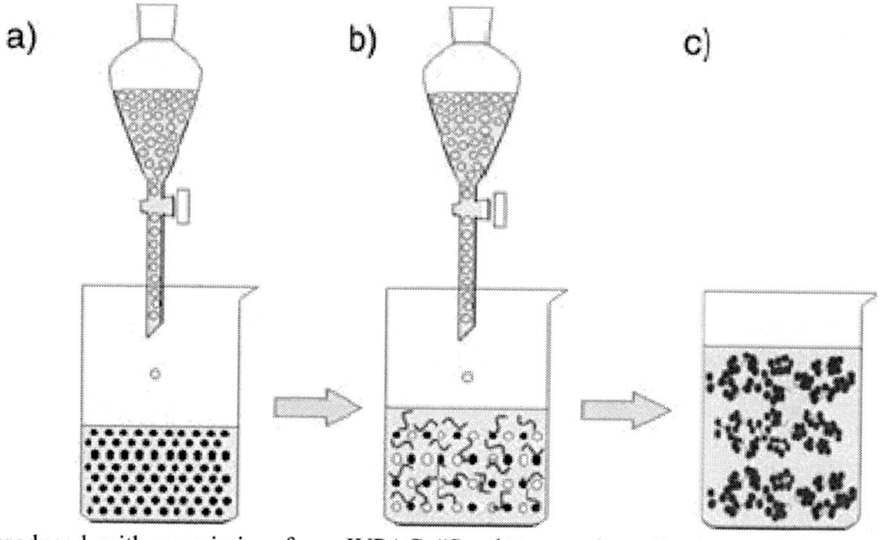

Reproduced with permission from IUPAC "Syntheses and applications of conducting polymer polyaniline nanofibers" J. Huang, Pure Appl. Chem., 2006, 78(1), 15–27.

Figure 10. Schematic diagram illustrating the formation of polyaniline agglomerates during conventional chemical synthesis. a) The oxidant (open circles) solution is added slowly to the aniline (solid circles) solution. b) Polyaniline nanofibers form as soon as the polymerization begins. Since the nanofibers are exposed to aniline and oxidant, they are subject to secondary growth. c) Agglomerates of polyaniline particles plus a few nanofibers are found because of severe secondary growth.

Interfacial Polymerization

Huang and Kaner [42] developed rapid mixing and interfacial polymerization method, Figure 12, to chemically synthesize polyaniline nanofibers.The interfacial polymerization represents one effective method to suppress secondary growth.

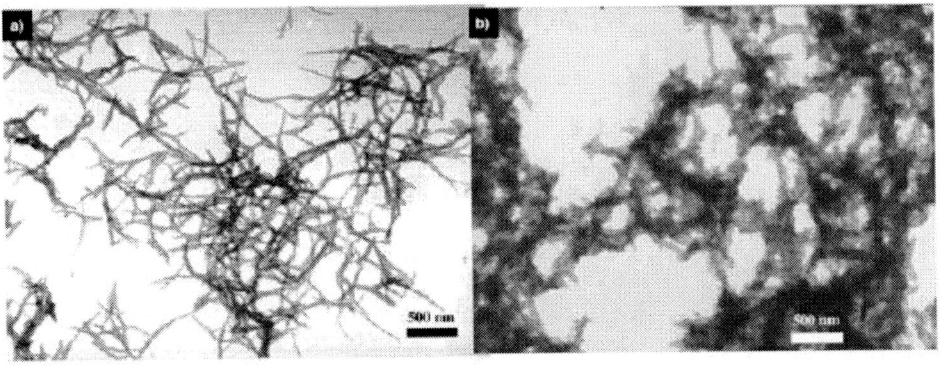

(a) (b)

Reproduced with permission form ACS "Polyaniline Nanofibers: Facile Synthesis and Chemical Sensor" J.Huang, S.Virji, B.H. Weiller, and R.B. Kaner, J.Am.Chem.Soc. 2003, 125, 314-315.

Figure 11. Typical TEM images showing the morphological evolution of polyaniline during chemical oxidative polymerization in 1m HCl at about 0^0C. The samples were extracted: a) as soon as the green color of polyaniline became visible, b) after 25 minutes.

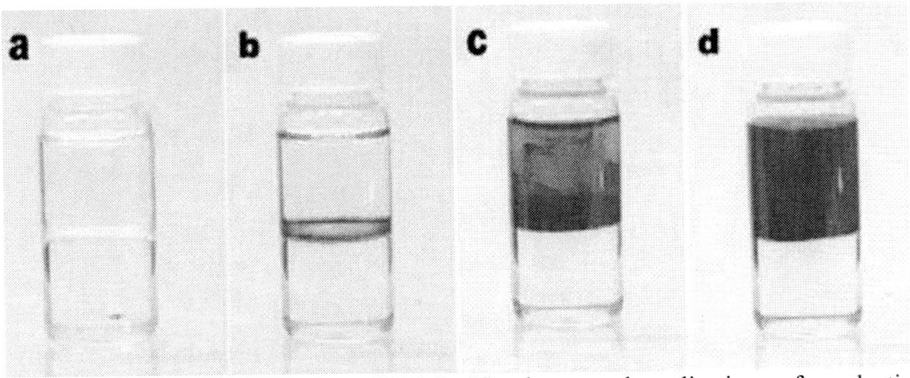

Reproduced with permission form IUPAC, "Syntheses and applications of conducting polymer polyaniline nanofibers" J.Huang, Pure Appl. Chem., 2006, 78(1) 15–27.

Figure 12. Pure polyaniline nanofibers can be made by interfacial polymerization. In a typical reaction, (a) an aqueous solution of acid (1 M HCl) and oxidant and an organic solution of aniline are brought together to form an interface. (b) Polyaniline first forms at the interface and then (c, d) diffuses into the water layer since as-prepared polyaniline is in its hydrophilic emeraldine salt form.

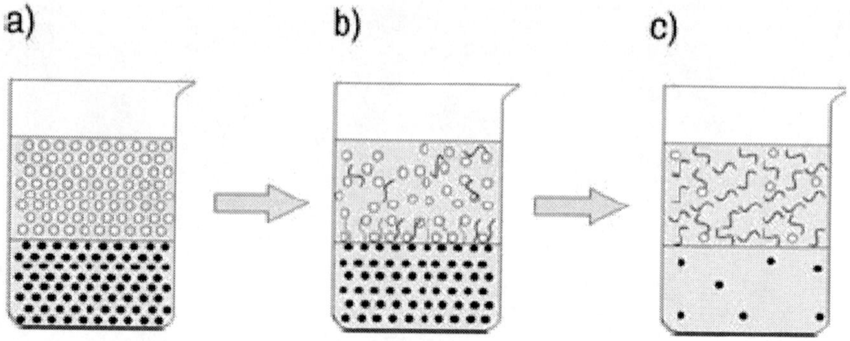

Figure 13. schematic presentation of the synthesis of polyaniline nanofibers by using interfacial polymerization. (a) An interface is set up between an organic phase containing dissolved aniline (solid circles) and an aqueous phase containing the oxidant (open circles). (b) Polyaniline nanofibers form at the interface, where aniline meets the oxidant and diffuse into the water phase. Note that this carries the nanofibers away from the reactive interface so they are not subject to further secondary growth. (c) As the polymerization proceeds, nanofibers accumulate in the aqueous phase.

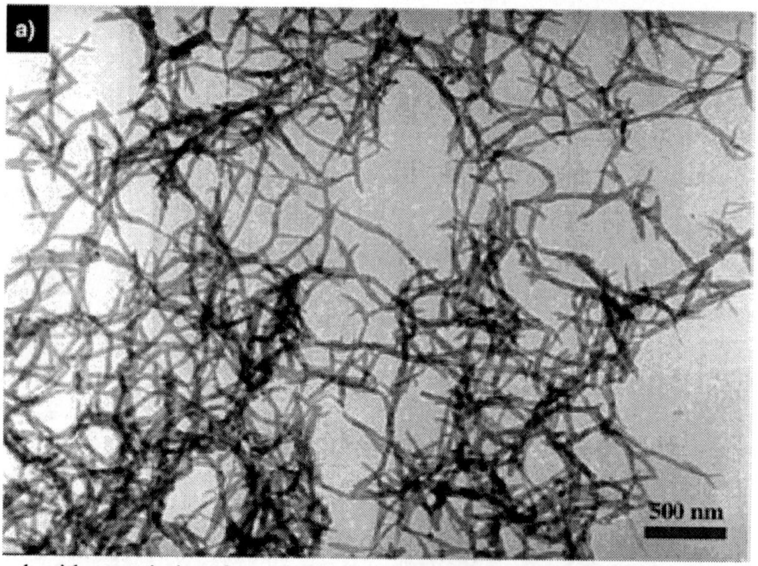

Reproduced with permission form IUPAC , "Syntheses and applications of conducting polymer polyaniline nanofibers" J.Huang, Pure Appl. Chem., 2006, 78(1) 15–27.

Figure 14. Typical TEM images showing the morphology of polyaniline synthesized by interfacial polymerization.

Since the monomer aniline and the initiator ammonium peroxydisulfate are separated by the boundary between the aqueous and the organic phase, polymerization occurs only at this interface where all the components needed for polymerization come together.Polyaniline then forms as nanofibers. Since these newly formed nanofibers are in the doped emeraldine salt form, they are hydrophilic and can rapidly move away from the interface and diffuse into the water layer Figure 13. In this way, as the nanofibers form they are continuously withdrawn from the reaction front, thus avoiding secondary growth and allowing new nanofibers to grow at this interface, Figure 14.

4.1.3. Template Synthesis

By introducing “structural directing agents” during the chemical polymerizing reaction, which include: surfactants [43-45], liquid crystals [46], nanowire seeds [47], aniline oligmers [48], and relatively complex, bulky organic dopants [49-52]. It is believed that such functional molecules can either directly act as templates (e.g., poly electrolytes) or promote the self-assembly of ordered “soft templates” (e.g., micelles, emulsions) that guide the formation of polyaniline nanostructures. Lately, Zhang et al. [47] have approached to well-designed conducting polymer nano-fibers by seeding route. Probably the most useful feature of template synthesis is that it is a general method with respect to the types of materials that can be prepared as well as templates that can be involved. Depending on the properties of monomers to be polymerized, the chemistry of the template, and their mutual interactions, e.g., whether the monomer interacts electrostatically or whether it is chemically bound to the template, different micro and nanostructures can be formed

The templates technique can be classified as

1. Hard templates
2. Soft templates
3. Seed templates

1. Hard Templates

The hard template approach, proposed by Martin [53-54], involves a synthesis of CP within the pores or channels of membranes. In general, two types of membranes were used by Martin’s group: track-etch polycarbonate (PC) or polyester (PE) membranes and anodic aluminum oxide (AAO) membranes.

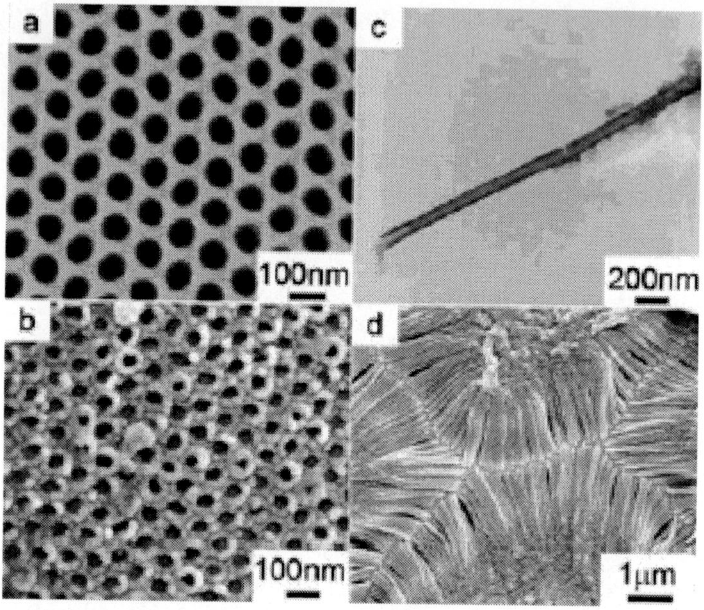

Reproduced with permission from ACS "Controlled Electrochemical Synthesis of Conductive Polymer Nanotube Structures" R.Xiao, S.Cho, R.Liu, and S.B.Lee, J.Am.Chem.Soc, 2007, 129, 4483-4489.

Figure 15. SEM images of (a) a homemade alumina template with 80-nm pores in diameter, (b) annular Au electrodes after the removal of alumina template, and (c) TEM and (d) SEM images of PEDOT nanotube synthesized at 1.2 V for 100 s in 10 mM EDOT.

Some other solid porous materials, such as zeolites [55,19], silica-based mesoporous molecular sieves[56], oxides [57-58,21-22], graphite composites [59-6023, 24],polyoxometallates [61], carbon nanotubes and Ag, Au, or Ni nanoparticles [62-65] can be used in hard template synthesis of CP nanostructures .The "*Track-etch films*" of hard templates are made by bombarding the material with high energy beams to create damage tracks in the material and then chemically etching these tracks into pores. The deposition process of CP within pores is accomplished by coating one surface of a membrane with a thin metal film, for example Pt or Au. Most of the studies in hard template synthesis have entailed the use of two types of nanoporous materials: particle track- etched polymeric membranes (PTMs) and porous alumina membranes obtained by anodic oxidation of aluminum in acid solution. PTMs are available with a wide range of pores diameters (some as small as 10 nm) and pores densities ranging from 10^5 to 10^9 pores per cm^2 whereas in the case of the alumina membranes, Figure 15 , the pore densities may be as high as 10^{11} pores per cm^2. Moreover, the structure of the polymeric and oxide membranes are different. Polymeric membranes, mostly PC or PE, contain randomly distributed cylindrical pores of uniform diameters

They are commercially available as microporous and nanoporous polymeric filtration membranes, with pore diameters ranging from 10 nm to several microns. The use of porous membranes with uniform pores yields monodisperse nanocylinders or nanorods, whose dimensions are carefully controlled by the pore size. These tubular or fibrillar structures can remain within the template or they can be extracted from the membrane. The porous materials in which polymer structures are grown can be considered as "encapsulating" materials aimed at the isolation of the functional, conducting micro- or nanostructure from the surroundings. Monomolecular layers provide an opportunity to define their chemical functionality and applications with molecular precision and become increasingly important tools in the design of surface-confined synthetic routes to complex structures; among such structures the conjugated polymers playa significant role. The use of self-assembled monolayers with suitable functional groups protruding towards the monomer-containing solution can lead - via the appropriate synthetic route - to covalent or electrostatic attachment of monomers to these functionalities.Subsequent polymerization results in 1-D or 2-D polymeric patterns (polymer brushes, "wires", surfaces), depending on the polymerization method [66-68]. By introducing "defect sites" in the self-assembled template monolayers or non-conducting polymers, through, e.g., micro contact printing (μCP) on the solid surface, region-specific electro-deposition of conducting polymer micro- and nano dots can be achieved. The disadvantages of the hard template methods are generally tedious post synthetic processes, which are required in order to remove the templates and the destruction of nanostructures during the post-synthetic process i.e. the formation of undesirable aggregated structures after removal of the templates.

2. Soft Templates

The soft template methods are based on the use of structure-directing molecules, such as various soluble oligomers and polymers, surfactants and amphiphilic acids which are able to form, alone or with aniline, aggregates like cylindrical micelles, and other supra molecular 1-D aggregates. Nanostructured conducting polymers have been prepared by oxidative polymerization of aniline in the presence of nonionic surfactants (Triton X-100) [69] anionic surfactants (sodium dodecylsulfate) [70] cationic surfactants (cetyltrimethyl ammonium bromide) [71] mixture of ionic surfactants (cetyltrimethyl ammonium bromide) and (sodium dodecylbenzene sulfonate) [72] amphiphilic ionic liquid (1-

hexadecyl-3-methylimidazolium chloride) [73] and amphiphilic acids e.g.,4-{n-[4-(4-nitrophenylazo)-phenyloxy] propyl , amino benzenesulfonic acid (ABSA) where n = 2, 3, 4, 6, 8, or 10, 77 ,78 azobenzene sulfonic acid [74], naphthalene sulfonic acid (NSA) [75] dodecyl benzene sulfonic acid (DBSA) [76] *p*-aminobenzene sulfonic acid (p-ABSA),[77] *p*-toluene sulfonic acid (PTSA) and camphor suphonic acid (CSA) [78], 4-[4-hydroxy-2-pentadec-8-enyl)phenylazo] benzenesulfonic acid ,[79] Allura Red AC,[80], cardanol ,[81] and sulphonated dendrons [82].

The use of the mixture of surfactants (N-cetyl N,N,N-trimethyl ammonium bromide) and acids (oxalic acid) is known to be effective, especially in the production of 3-D conducting polymer network. The formation of nano conducting polymers in the presence of amphiphilic acids is found to depend mostly on the initial acid/aniline ratio, and good morphology of the nanofibers was produced only for selected compositions. The reason for the in homogeneity in the morphology was associated to the poor and unstable micelles formation of aniline-acid complex in water.

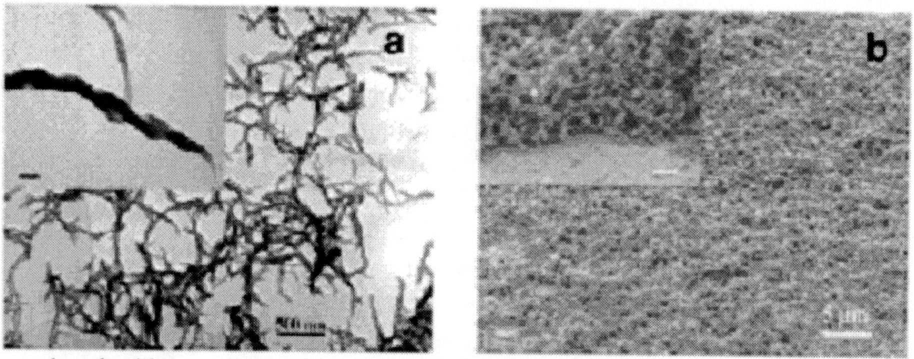

Reproduced with permission form ACS "Polyaniline Nanofibers: Facile Synthesis and Chemical Sensor" J.Huang, S.Virji, B.H. Weiller, and R.B. Kaner,, J.Am.Chem.Soc., 2003, 125, 314-315.

Figure 16. TEM images of polyaniline/CSA nanofibers cast from a suspension after dialysis. The inset shows a twisted fiber (scale bar) 50nm). (b) SEM secondary electron images of a thin film deposited on glass from the suspension. The inset shows a cross-sectional view of the film on the glass substrate (scale bar) 200 nm).

It should be noted here that micelles of amphiphilic acids and/or corresponding aniline salts were tentatively proposed to interpret the formation of nanostructured PANI and other nanostructures, Figure 16. However, the existence of cylindrical micelles has never been unambiguously proved for the investigated amphiphilic acid/aniline systems. Nanofibers of self-doped PANI were synthesized by the oxidation of aniline with APS in an aqueous solution, in the presence of *o*-aminobenzenesulfonic acid, and *m*-aminobenzenesulfonic acid [83].Helical PANI nanostructures were prepared by using the chiral CSA as a dopant [78]. Chiral PANI nanorods and gold microspheres were obtained by the oxidative polymerization of aniline with potassium tetrachloroaurate (III) in the presence of (+)-CSA.PANI hollow microspheres containing PANI (20–30 nm in diameter) were prepared by inverse microemulsion polymerization in the presence of NSA as the dopant.

3. Seed Template Method

Seeding an ordinary chemical oxidative polymerization with very small amounts (usually <1%) of nanofibers, regardless of their chemical nature (inorganic or organic), and single-walled carbon nanotube (SWNT) bundles. The morphology of the nanofibrillar (particulate) PANI emeraldine salt precipitate quantitatively to nanofibers having an average diameter in the range 20–60 nm. The general shape of the seed (nanofibers vs nanospheres) controls the overall morphology of the resulting precipitate while differences in the length, diameter, etc., of the nanostructured seed templates do not have a substantial impact. The in-situ deposited nano conducting polymeric films on the walls of the reaction flasks, in reaction mixtures seeded with SWNT,yields thin (<1 μm), transparent, nanofibrillar which are strongly adherent. It has been revealed that the stirring of reaction solution has a dramatic effect on the morphology of conducting polymers synthesized by the nanostructured seed template method. If no stirring is used, a large amount of new nanofibers appear in the product, while the morphological change of the pre-added fibers is unremarkable. In contrast, if the reaction is stirred, granular nanoparticles will gradually grow on the pre-added fibers and the surface roughness as well as thickness of the pre-added fibers increases with the addition of the oxidant, leading to the formation of coral-like fibers.Nanoconducitng polymers have been synthesized using SWNT, chlorophyll in nanorods, and composite nanofibers of methyl orange/ferric chloride, as nanostructured seed template, Figure 17 [84].

It has been observed that the conducting polymer nanofibers prepared in different organic solvents as seed solution yields uniform nanofibers with relatively high conductivity,Figure 18 [85] . The accelerated introduction time and the seeded growth are probably responsible for the formation of the nanofibers. The interaction between the solvents and the monomers influences the structure, conductivity and water dispersity of the resulting nano conducting polymers to different extents.

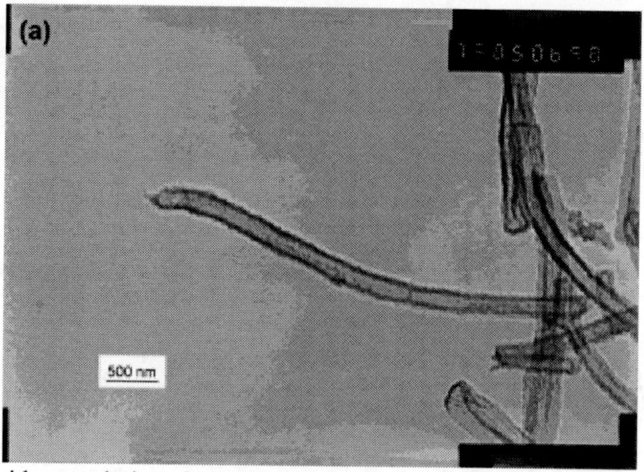

Reproduced with permission form Elsevier "Electrochemical synthesis of functional polypyrrole nanotubes via a self-assembly process" X.Yang , T.Dai , Z.Zhu , Y.Lu, Polymer, 2007, 48, 4021-4027.

Figure 17. TEM image of carbon nanotubes obtained from PPy/MO.

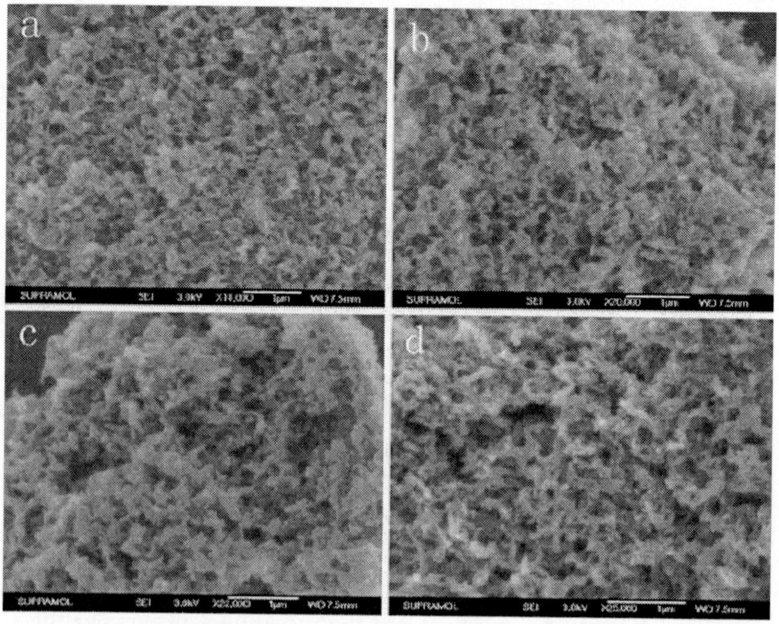

Figure 18. SEM images of the samples using different kinds of organic solvents for the polymerization (a-DMF; b-THF; c-DMSO; d-NMP).

The stronger the interaction is, poorer water dispersity is shown by the polymer. This seed polymerization method allows the synthesis of nanostructured conducting polymers with higher conductivity thereby widening the scope of ease of nano synthesis of these materials.

4.1.4. Biocatalytic Approach

Biocatalytic polymerization using a naturally occurring enzyme is advantageous as it is a one-step, environmentally compatible synthesis with the potential for producing industrial polymers in high yield due to the high efficiency of the biocatalyst. This technique has been successfully employed as a catalyst for polymerization of phenols and anilines [86-93].The free radicals of the monomer (substrate) in aqueous solution undergo oxidative coupling to produce dimers and other oligomeric products. The successive oxidation and coupling of oligomers eventually result in the formation of high-molecular-weight polymers [89]. This biocatalytic approach has been developed to synthesize water-soluble conducting polyaniline. The mechanistic cycle of *horseradish peroxidase* (HRP) involves a two-electron oxidation step and two one-electron reduction steps [94],Figure 19. Recently, the synthesis of conducting nanocomposites guided by the PAA template was studied, suggesting that HRP influences directly the stereo-specificity of the PANI in the nanocomposites, PAA/PANI/(+)CSA, and PAA/PANI/(–)CSA,Figure 20 [90-91].HRP appears to be a suitable enzyme for the synthesis of such polymers , although other *peroxidases* such as that derived from soybeans are viable alternatives [94]. HRP is able to catalyze the oxidation of a wide range of compounds, including aromatic amines and phenols, in the presence of hydrogen peroxide to generate corresponding free radicals

$$\text{HRP} + H_2O_2 \rightarrow \text{HRP I}$$

$$\text{HRP I} + \text{RH} \rightarrow \text{R*} + \text{HRP II}$$

$$\text{HRP II} + \text{RH} \rightarrow \text{R*} + \text{HRP}$$

The substrate in this case (RH) is aniline. R* is the radical species formed from aromatic amine. These free radicals then undergo coupling to produce dimer, and successive oxidation and coupling reactions eventually result in the formation of polymer [95-96]. The enzymatic approach is *environmentally benign*, offers a higher degree of control over the kinetics of the reaction, and has the potential of producing products in high yields, Figure 19.

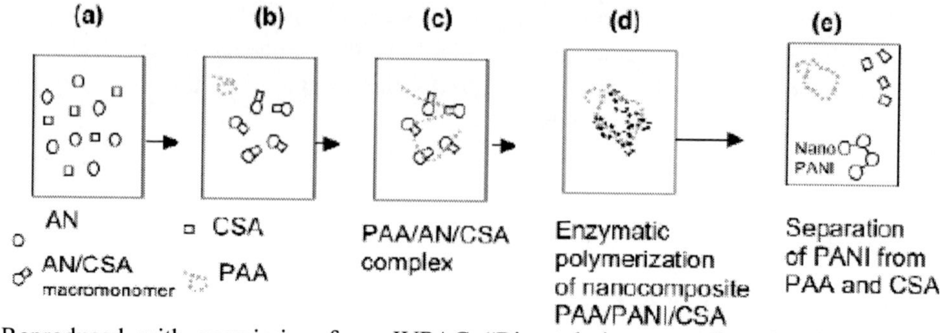

Reproduced with permission from IUPAC "Biocatalytic approaches for synthesis of conducting polyaniline nanoparticles", A.L.Cholli, M.Thiyagarajan, J.Kumar, and V.S. Parmar, Pure Appl. Chem., 2005,77(1),339–344.

Figure 19. Scheme showing the steps (a–e) that are involved in the enzymatic polymerization of nanocomposite PAA/PANI/CSA possessing a specific helical conformation and PANI nanoparticles.

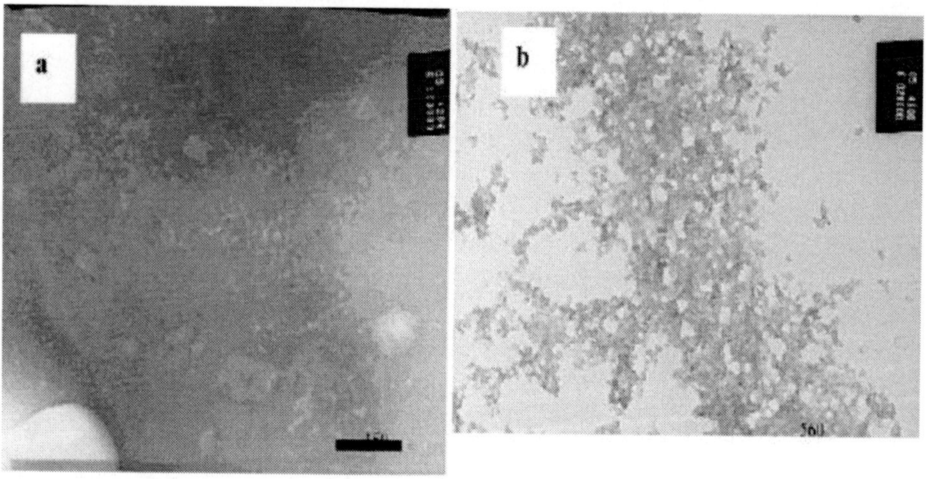

Reproduced with permission from IUPAC "Biocatalytic approaches for synthesis of conducting polyaniline nanoparticles",A.L.Cholli, M.Thiyagarajan, J.Kumar, and V.S. Parmar, Pure Appl. Chem., 2005,77(1), 339–344.

Figure 20. TEM images of (a) conducting nanocomposite PAA/PANI/(–) CSA, (b) after treating with NH_4OH.

4.1.5. Chemical Vapor Deposition

The thermal chemical vapor deposition polymerization (CVDP) of the conjugated polymers demonstrates that CVDP is quite an effective top-down method to fabricate nano-materials, nano-composites, and conformal coating on nano-structures [97]. It is a great challenge to prepare the pristine form of most nano conducting polymers in desired shapes and dimensions. This task is often accomplished by first synthesizing soluble derivatives of the pristine polymer, dissolving it in solvents, and preparing polymer films via solution-based processes including simple solution casting, spin coating, and screen printing. Investigations were directed to realize ordered bulk nano materials by polymer solution or melted polymer. As CVDP depends on the solution methods to fabricate electronic and optoelectronic devices based on conducting polymers at nano scale, it provides a new approach for the above mentioned applications.

4.2. Classification of Experimental Procedures

Availability of powerful nano characterization tools such as HR-TEM, and AFM, have made possible to give insights into different levels of crystallite and agglomerate sizes. The size control measurement method is necessary for obtaining information about the applicational potential of nano conducting polymers .Hence nano characterization techniques provides insight into primary nanoparticle morphology comprising structure , defects, impurity, surface characteristics, interfaces and boundary defects as well as end use applicational properties such as conductivity, mechanical characteristics etc.

4.3. Structural Characterization

4.3.1. FT-IR

The FT-IR spectra of the nano PANI synthesized in different media are shown in Figure 21. The presence of the characteristic peaks of PANI doped with HCl indicates the successful polymerization of aniline [85]. The peaks at ca. 1560 and 1490 cm^{-1} are assigned to C=C stretching of the quinoid and benzenoid rings, respectively, and the peak at ca. 1140 cm^{-1} represents the C-H aromatic in-plane bending. The peak observed at 1292 cm^{-1} is related to the C-N stretching vibration of the benzenoid ring. The band falling in the range 800-860 cm^{-1} is identified with the out-of-plane bending of C-H bond in the 1,4-disubstitued ring, which has been used as a key to identify this type of substituted benzene. The peak at about 1140 cm^{-1} is often related to the doped structure; therefore, the results suggest that the PANI obtained by using aniline seed dissolved in organic solvents is more protonated than that prepared from the conventional polymerization, Figure 21. The peak normally observed at 1136 cm^{-1} undergo red-shift to about 1128 cm^{-1} when using the seed polymerization with aniline/DMF. The other samples also exhibited different degrees of red-shifts, indicating that a slightly higher doping level was obtained when the aniline was polymerized using the seed method.

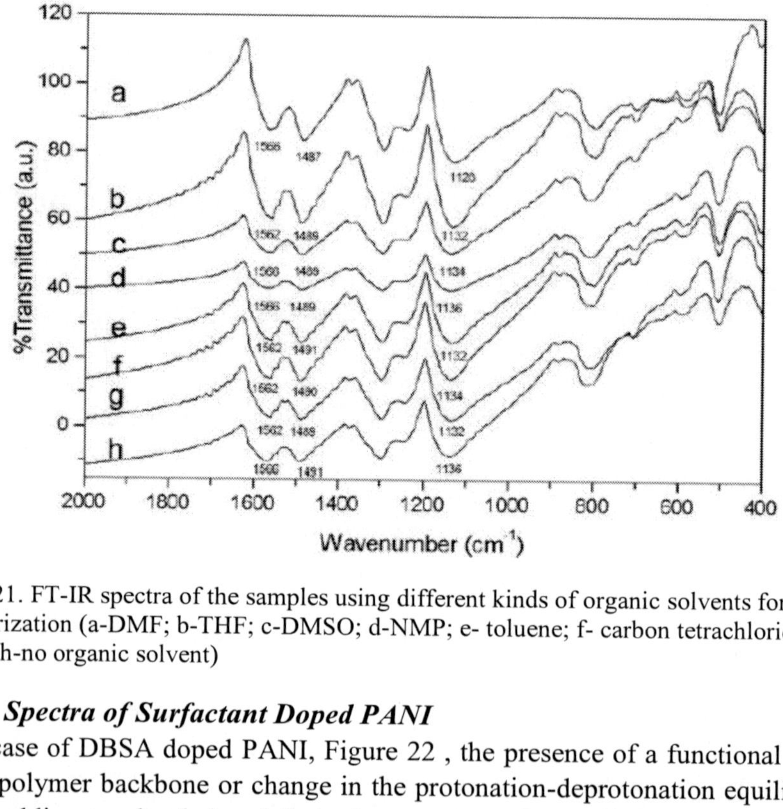

Figure 21. FT-IR spectra of the samples using different kinds of organic solvents for the polymerization (a-DMF; b-THF; c-DMSO; d-NMP; e- toluene; f- carbon tetrachloride; g-xylene; h-no organic solvent)

FT-IR Spectra of Surfactant Doped PANI

Incase of DBSA doped PANI, Figure 22 , the presence of a functional group on the polymer backbone or change in the protonation-deprotonation equilibrium of emeraldine can be deduced from the presence of corresponding bands in the FTIR spectrum[98].The presence of the imine and amine bands is an indication of the polyaniline in emeraldine state (EB). The vibrational frequencies specific for EB are 1560, 1477, 1399, 1122 and 827 cm^{-1}.The spectrum represents benzenoid and quinoid ring vibrations at 1477, 1560cm^{-1}. The ratio of the absorption heights of quinoid to benzenoid ring modes (1560-1477) suggests that the percentage of quinoid units is almost equal to benzenoid units in EB and low in ES. The results show that obtained polyaniline contain 1,4-para-disubstituted linear chains[98].

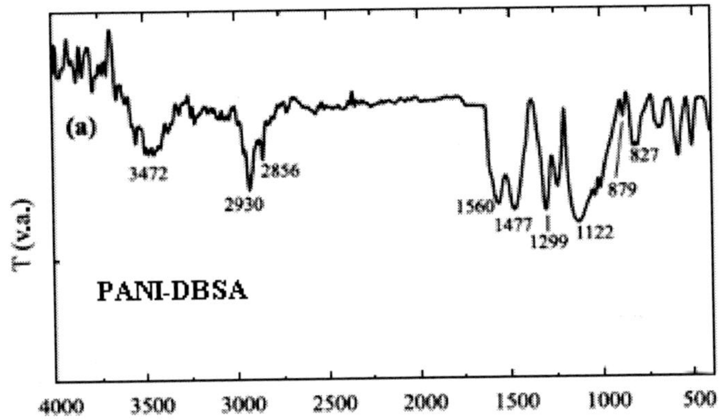

Figure 22. FT-IR spectra of DBSA doped PANI.

Raman spectroscopy has been used to characterize the structure of nano conducting polymers such as PANI. The major bands observed in nanostructured polyaniline are given in Table1. The resonance enhancement depends on the concentration of radical cations in the dispersion.[99].The Raman spectra of PANI samples synthesized in different media are presented in Figure 23. The bands at about 1600 and 1400 cm^{-1} assigned to the C-C stretching vibration of the benzenoid and quinoid rings can be observed. The bands at ca. 1500, 1335 and 1167 cm^{-1} corresponding to the C=N stretching of the quinoid diimine units, the C-N stretching of the cation radical and the C-H bending vibration, respectively, are also found. The bands at about 800 and 417 cm^{-1} are related to the C-H deformation, the band at 575 cm^{-1} is attributed to the amine deformation and the band at 525 cm^{-1} can be ascribed to the C-N-C torsion [100-102]. When the organic solvents are added into the reaction systems, the conjugation length of the PANI is extended or the PANI owned a higher doping level [103].The presence of strong bands that are mainly due to benzenoid segments at lower pH, confirming that the radical cations involved are polarons and not bipolarons. An explanation for the change in intensity of the Raman bands at higher pH (> 8) is complicated because in this pH range Raman band intensity is controlled by coagulation of PANI particles and concentration of quinonoid segments.

Table 1.Characterisitcs peaks of polyaniline observed in Raman Spectra

Frequency (cm^{-1})	State*	Assignment [90, 91]
1627–1637	B	C–C ring stretching
1590–1597	Q	C–C ring stretching
1540–1546	Q	C–C ring stretching
1506–1516	Q	C=N stretching
1490–1496	Q	C=N stretching
1230–1235	Q	In-plane ring deformation
1200–1205	B	C–H in-plane bending
1171–1177	Q	C–H in-plane bending
890–895	B	In-plane ring deformation
845–855	B	Amine ring deformation
825–830	Q	In-plane ring deformation
800–815	Q	C–H out-of-plane bending
712–724	B	Out-of-plane ring deformation
685–698	Q	Out-of-plane ring deformation
640–645	B	In-plane ring deformation
511–526		Out-of-plane C–N–C torsion
412–420	Q	Out-of-plane C–H wag

*: B = benzoid ring and Q = quinoid ring

4.3.2. Raman Spectroscopy

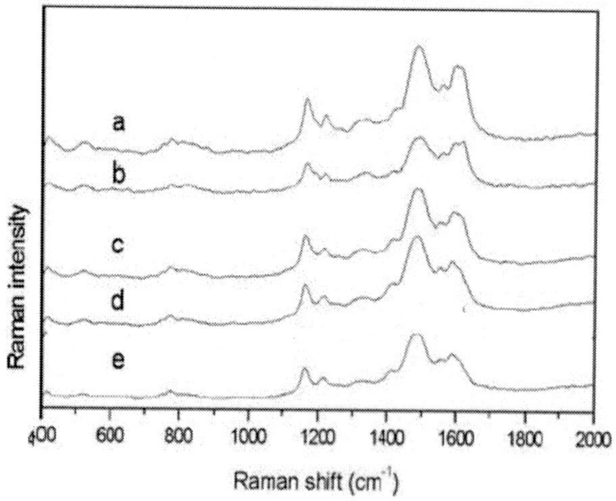

Figure 23. Raman spectra of PANI obtained from different systems (a. DMSO, 1 ml; b. DMF, 1 ml; c. THF, 1 ml; d. NMP, 1 ml; e. no organic solvent).

Raman Spectra of Surfactant Doped PANI

The PANI-β-NSA spectra with different β-NSA/Aniline molar ratios are given in Figure 24. The doping is confirmed by the presence of the bands at 1167 and ca.1310-1332 cm^{-1}, assigned to the radical cation segments (1165 and 1320-1336 cm^{-1} for PANI-ES spectrum. The enhancement of the relative intensities of the bands at 1508 and 1592 cm^{-1} appears to be striking .There are two assignment for these bands: the bands can be associated to n N-H and n C-C of quinoid units, such as in ES form (1515 and 1578 cm^{-1} in the PANI-ES spectrum) or they can be assigned to n C=N and n C-C of quinoid units, such as in emeraldine base/ EB form (1470 and 1585 cm^{-1} in the PANI-EB spectrum, Figure 24.

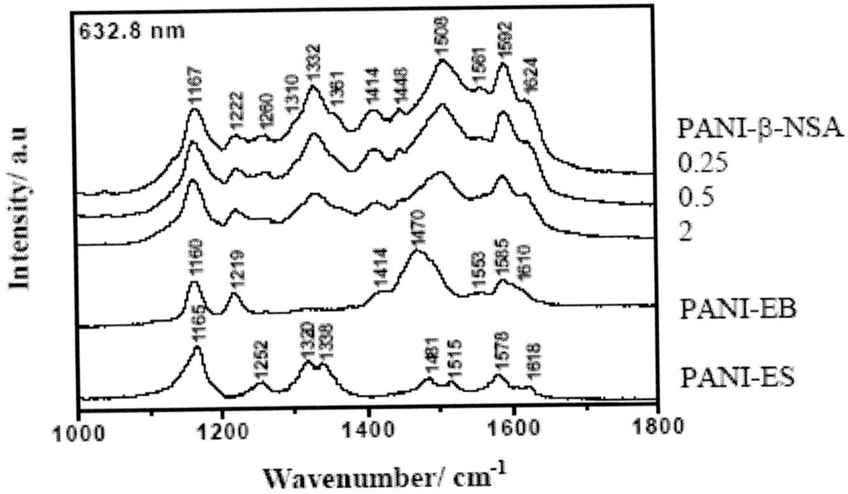

Figure 24. Raman spectra of powder samples: PANI-ES and PANI-EB prepared and PANI-β-NSA synthesized with different b-NSA/Aniline molar ratio (indicated in the figure).

The UV-VIS-NIR spectra of PANI- β-NSA (not shown) are similar to that of PANI-ES and the relative intensity of the bands at ca. 1310-1332, to those at 1508, and 1592 cm^{-1} increases with the decreases of β-NSA/Aniline molar ratio used in the synthesis of PANI-β-NSA. These results preclude that the bands at 1508 and 1592 cm^{-1} could be attributed to quinoid units such as in the PANI-EB[104].

4.3.3. UV-Vis Spectra

A typical absorption spectrum of PANI has two absorption peaks at about 330 nm and 420 nm and one absorption peak λ_{max} at about 850 nm [99]. The first two peaks often combine to emerge as a platform. The peak at 330 nm arises from π-π* electron transition within benzenoid segments and the peak at 420 nm, accompanied by the absorption with maximum of 850 nm is related to the doping level and formation of polaron [99].

UV-visible Spectra of Surfactant Doped PANI

The spectra of the polyanilines are very complex owing to multiple individual distribution of the quinoid and benzenoid rings along the polyaniline backbones, different final groups, minor substituents, different products resulted as secondary possible reactions, cyclizations, conformation effects and reduction in ring symmetry [105-106].The induction period depends largely on reaction parameters: it is longer at low temperature and at higher molar ratio of monomer/oxidant. After the addition of 3/4th of the ammonium per sulfate, the reaction mixture turns from clear transparent to light blue.

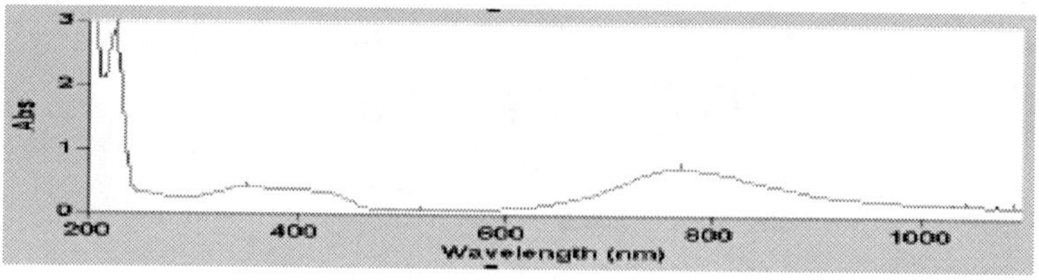

Figure 25. UV visible spectra of DBSA doped PANI.

After the addition of all the oxidant, the mixture became dark blue and during reaction (3h), dark green. The band at 740 nm, Figure 25 (originating from polaron transitions) is an indicator of the progress of the reaction. Initially, it shows a bathochromic shift indicating an increase in conjugation length of the polymer backbone and a high protonation.

The band in the region of 730–795 nm attains a constant position (770 nm) approximately 3 h after the addition of AP, indicating the completion of the polymerization. Electronic absorption spectra show three bands: 750–770 nm (the transition of electron from benzenoid ring to quinoid ring), 430 nm (π-π* transition in polaron/bipolaron) and 350-360 nm (π-π* transition of the benzenoid rings) indicating that the polymer is in its emeraldine salt form [106].The band at 770 nm is sensitive to changes in the concentration of the dopant. Normally, an increase of the DBSA concentration decreases the pH of the medium thereby decreases the band gap due to higher protonation. Yin and *Ruckenstein* [106] suggested that an increase in the amount of dopant in the polymer increases the compact coil conformation of the polymer, leading to a hypsochromic shift. The pH influences the protonation-deprotonation equilibrium of conducting emeraldine salt and non-conducting emeraldine base forms of polyaniline. Emeraldine base has two bands at 550–600 (n-π^* transitions - quinone-imine groups-exciton transition of the quinoid rings), and 315–330nm (π-π^* transitions) [107].The position and the absorbance of the band at 350 nm are unaffected when the pH < 8.Further increases in the pH increases the absorbance of this band and causes a blue shift because of the emeraldine salt-to-emeraldine base transformation. The band at 420 nm, assigned to the polaron transition, disappears for pH > 9. The decrease in the absorbance of the band situated around 775nm as the pH increases is attributed to the decrease in protonation of the polymer backbone which in turn decreases the number of polarons.At pH >9 an increase of the conjugation length and a decrease of the band gap takes place. This demonstrates that in this pH range, the transformation of protonated benzenoid structures into unprotonated quinonoid structures occurs [98,107].

4.3.4. X-ray Diffraction

The X-Ray diffraction (XRD) patterns of nanostructured undoped PANI and PANI are shown in Figure 26. The XRD of undoped PANI clearly suggests that it is amorphous. The appearance of a large number of sharp diffraction peaks upon doping with HCl, suggests significant crystallization of PANI on protonation. The evolution of crystalline islands upon protonation has been directly correlated with the enhancement of conductivity [108].

4.3.5. Morphological Properties

4.3.5.1. SEM Analysis

It is well established that morphology of the nanostructures synthesized electro-chemically or chemically strongly influences the properties. The wide range of SEM magnification makes it suitable for investigation of microstructures and sometimes nanostructures of PANI. With the help of SEM images, one can correlate the electrical conductivity, crystallinity, mechanical properties, with the structure of the nanoparticles. The SEM images of conductive polymer nanotubes are quite different from those of metallic nanotubes because conductive polymers are not as rigid as metals. Thus, it is important to characterize correctly three different nanostructures of conductive polymer nanotubes (nanowire, partially filled nanotube, and complete nanotube) by using SEM and TEM.

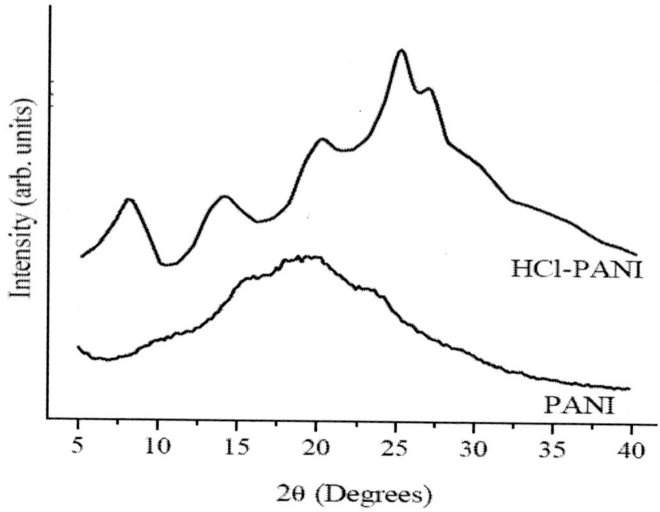

Figure 26. X-ray spectra of PANI and doped PANI.

4.3.5.2. TEM

The high resolution of the transmission electron microscope makes it a useful technique for determination of nanostructures. Conducting polymers have significant contribution in the area nanotechnology as active components of nano molecular electronic devices. Investigation of nanostructures is essential and can be done with transmission electron microscopy. The additional information gained from this technique is the distribution pattern of nanoparticles.

4.3.5.3. AFM

The AFM studies indicate the obtaining of a nanostructured material ,Figure 27.Statistical analysis of the AFM micrograph proves the existence of a non-normal distribution of the particle size,within a considerable range (7.5nm - 76.6nm) [109].

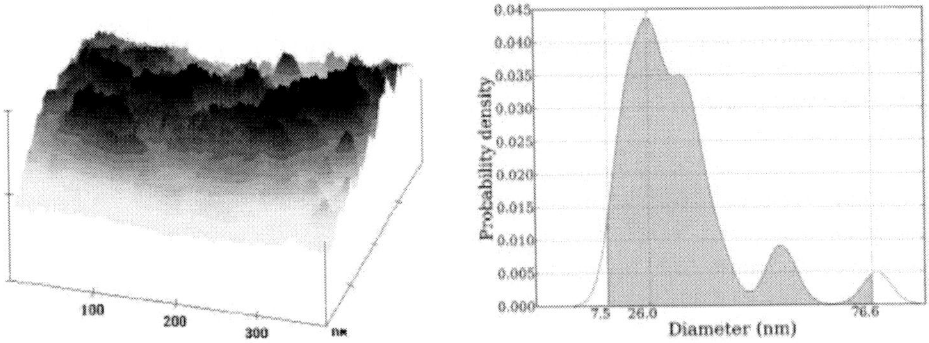

Figure 27. AFM of PANI, probability density of PANI nanoparticles.

4.3.6. Mechanism of Conduction

Figure 28 shows the chemical structure of several of the most important polymers which become electrically conductive when doped with oxidizing or reducing agents. All have conjugated double bonds along the main chain, and it is this chemical feature that makes conductivity possible. In the pristine state, conjugated polymers are semiconductors with band gaps between 1.5 and 3 eV. Oxidation or reduction either removes or adds electrons to the conjugated π system of electrons, leading to partially filled bands and therefore to metal- like conductivity.

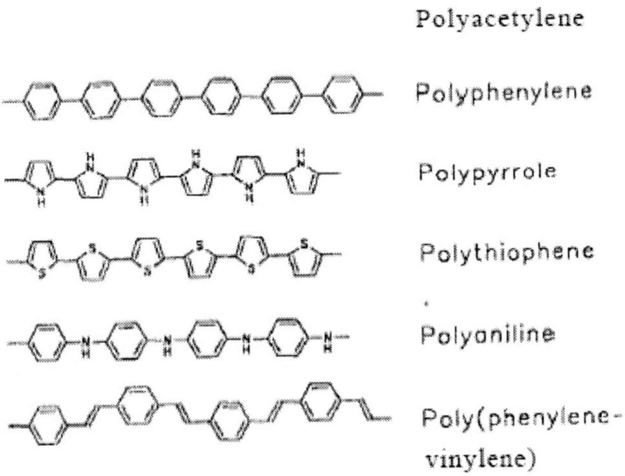

Figure 28. Chemical structure of some important conducting polymers.

In a polymer, just as in a crystal, the interaction of a polymer unit cell with all its neighbors leads to the formation of electronic bands. The highest occupied electronic levels constitute the valence band (VB) and the lowest unoccupied levels, the conduction band (CB). The width of the forbidden band, or band gap *(Eg),* between the VB and CB determines the intrinsic electrical properties of the material. For all the organic conjugated polymers, the band gap is larger than 1.5 eV so that these materials are intrinsically insulating. Initially, the high conductivity increase observed upon doping organic polymers was thought to result from the formation of unfilled electronic bands. Conducting polymers are usually polyconjugated structures which are insulators in the pure state but when treated with an oxidizing or a reducing agent can be converted into polymer salts with electrical conductivities comparable to metals. This is called doping which are generally of two types

- *P-type doping:* removal of electrons from the valence band by the oxidizing agent, leaving the polymer with a positive charge.
- *N-type doping:* it encompasses donation of an electron to the empty conduction band by a reducing agent. The doping reaction is actually a redox reaction.The first step is the formation of a cation (or anion) radical, which is called a soliton or a polaron.

$$P_n \Leftrightarrow [P_n^{+} A^{-}]$$

(reduction oxidation)

This step may then be followed by a second electron transfer with the formation of a dication (or dianion) known as a bipolaron.

$$[P_n^{+} A^{-}] \Leftrightarrow [P_n^{2+} 2A^{-}]$$

(reduction oxidation)

Alternatively after the first redox reaction, charge transfer complexes may form between charged and neutral segments of the polymer, when possible.

$$[P_n^{\bullet+} A^{-}] + P_m \rightarrow [(P_n P_m)^{\bullet+} A^{-}]$$

In p-type or n-type doping, electrons are , removed from the top of the valence band and in the n-type doping the electrons are added to the bottom of the conduction band, in analogy to the mechanism of generation of charge carriers in doped inorganic semiconductors. This assumption was however discarded by the discovery that polyacetylene (PA) polyparaphenylene (PPP), and polypyrrole (PPy) can display conductivity which does not seem to be associated with unpaired electrons but rather with spin less charge carriers, the entities that are responsible for conduction. In organic *molecules,* the equilibrium geometry in the ionized state is different from that in the ground state, e.g., the geometry of neutral biphenyl is benzenoid-like but becomes quinoid- like in the ionized state. The energies involved in the ionization process of a molecule are schematically depicted in Figure 29.

The geometry of the molecule is first distorted in the ground state in geometry of the ionized state. This distortion involves an elastic energy E_{dis} 30. If we consider the one-electron energy levels of the molecule, this distortion leads to an upward shift ΔE of the highest occupied molecular orbital (HOMO) and a downward shift of the lowest unoccupied molecular orbital (LUMO). The energetically favorable geometry relaxation in the ionized state is obtained when the quantity EIP-V - E1p-d (which actually corresponds to ΔE as can be inferred from Figure 30 is larger than the distortion energy Edis

In a *polymer chain,* considering the case of oxidation, i.e., the removal of an electron from the chain, we lower the ionization energy by an amount ΔE. If $\Delta E > E_{dis}$, we get a polaron.The polaron is just a radical ion (spin associated with a lattice distortion and the presence of localized electronic states in the gap referred to as polaron states. The ability of a charge to significantly deform the lattice around it is the manifestation of a strong electron-phonon coupling. The quantity ΔE - E_{dis} (E_{rel}) corresponds to the polaron binding energy. The calculations (based on Huckel theory have indicated that polaron formation is energetically favorable in all the organic conjugated polymers. When a second electron is removed from the polymer chain: we get a bipolaron which is defined as a pair of like charges associated with a strong local lattice distortion. The formation of a bipolaron implies that the energy gained by the interaction with the lattice is larger than the Coulomb repulsion between the two charges of same sign confined in the same location.

The electronic band structure corresponding to the presence of two polarons and that of one bipolaron is depicted in Figure 31. Since the lattice relaxation around two charges is stronger than around only one charge, E_{dis} for the bipolaron is larger than E_{dis} for the polaron and the electronic states appearing in the gap for a bipolaron are further away from the band edges than for a polaron.

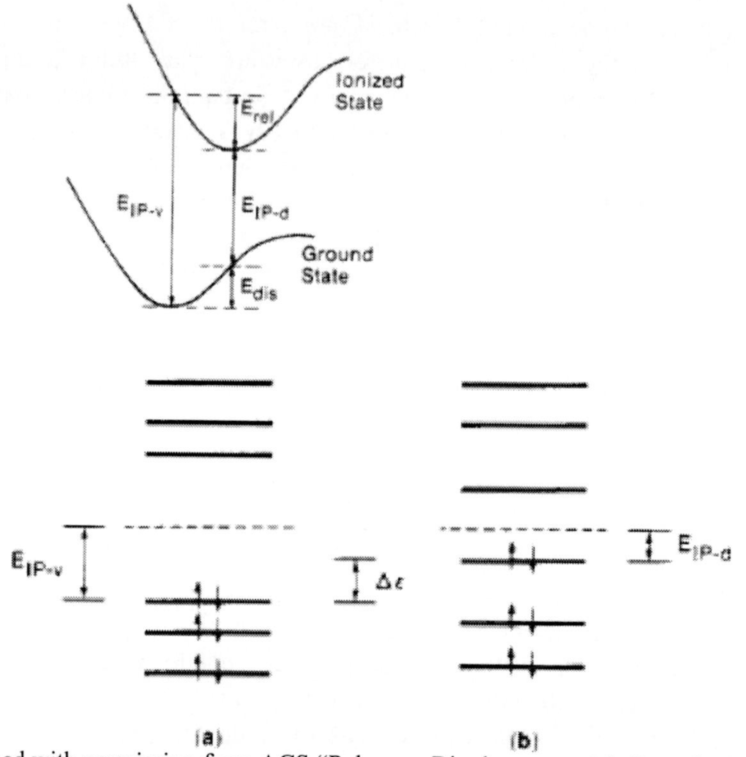

Reproduced with permission form ACS "Polarons, Bipolarons, and Solitons in Conducting Polymers" J.L.Bredas and G.B.Street , Acc. Chem. Res.,1985,18, 309-315.

Figure 29. Representation of the energies involved in a molecular ionization process. E_{IP+v} is the vertical ionization energy, E the relaxation energy gained in the ionized state, E_{dis} is the distortion energy to be paid in the ground state in order that the molecule adopts the equilibrium geometry of the ionized state, and E_{IP-d}, the ionization energy of the distorted molecule.(b) one-electron energy levels for an organic molecule (e.g., biphenyl) in its ground-state electronic configuration adopting: (a) the equilibrium geometry of the ground state; (b) the equilibrium geometry of the first ionized state.

The distortion energy E_{dis} to form one bipolaron is roughly equal to that to form two polarons while the decrease in ionization energy is much more in the bipolaron case than for two polarons. This is the reason why one bipolaron is thermodynamically more stable than two polarons despite the Coulomb repulsion between two similar charges. The bipolaron binding energy is calculated to be larger than that of two polarons In case of p-(n-) type doping, the bipolaron levels in the gap are empty (fully occupied), Figure 31.

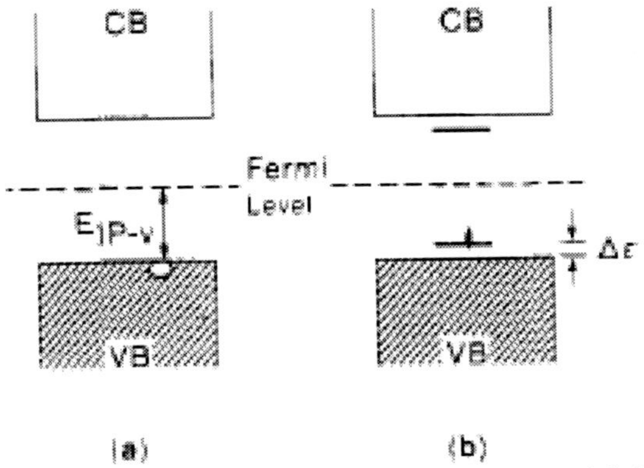

Reproduced with permission form ACS "Polarons, Bipolarons, and Solitons in Conducting Polymers" B.R.DAS and G.B.Street , Acc. Chem. Res.,1985,18, 309-315.

Figure 30. Band structure of a polymeric chain in the case of (a) a vertical ionization process and (b) the formation of a polaron. The chemical potential, or Fermi level, is taken as reference level.

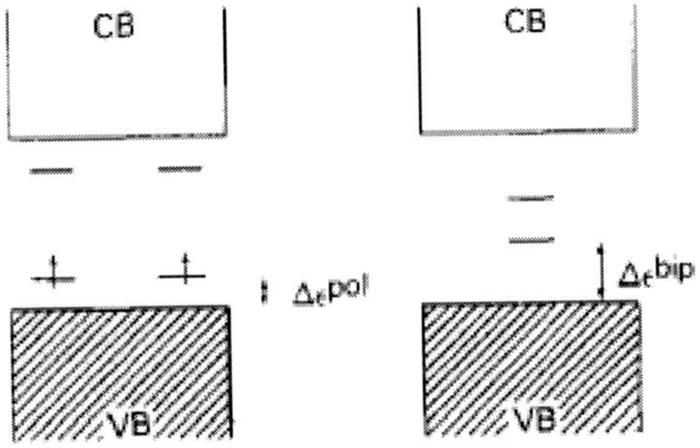

Reproduced with permission form ACS "Polarons, Bipolarons, and Solitons in Conducting Polymers" B.R.DAS and G.B.Street , Acc. Chem. Res.,1985,18, 309-315.

Figure 31. Electronic band structure a polymer chain containing: (a) two polarons; (b) one bipolaron.

The bipolarons are thus spin less. The presence of bipolarons on polymer chains result in the possibility of two optical transitions below the band gap transition: for p-type doping, from the VB to the lower bipolaron level and from the VB to the upper bipolaron level.

Table 3. Band structures of some conducting polymers

structure	polymer	π–π* bandgap (eV)	condcutivity (s/cm)
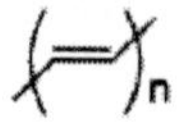	polyacetylene	1.5	10^3 - $1.7x10^5$
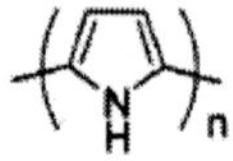	polypyrrole	3.1	10^2 - $7.5x10^3$
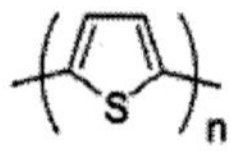	polythiophene	2.0	10-10^3
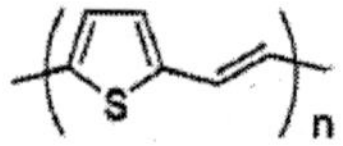	poly(thienylene vinylene)	1.6	40
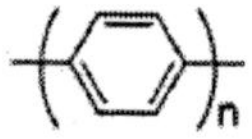	poly(*para*-phenylene)	3.0	10^2 -10^3
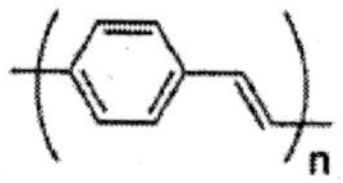	poly(*p*-phenylene vinylene)	2.5	3-$5x10^3$
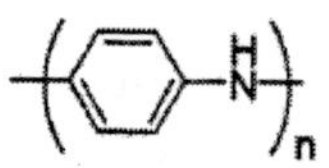	polyaniline	3.2	0-200

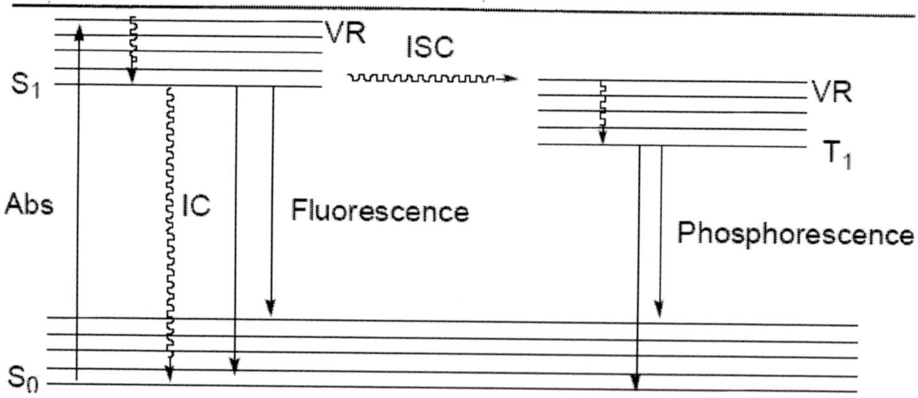

Figure 32. Emission mechanism.

4.3.7. Optoelectronic Properties

4.3.7.1. Luminescence

Generally, the observed transitions in absorption spectroscopy of conjugated polymers are attributed to electronic excitation from π to π* states, and π* to π states for emission spectroscopy. Upon electronic excitation of the polymer, a number of photo-physical processes, shown in Figure 32, may occur: fluorescence, phosphorescence, or radiation less decay [113]. Fluorescence is observed after singlet relaxation from the first excited state. If intersystem crossing occurs, a triplet excited state is generated whose relaxation will result in phosphorescence. If emission does not occur, then a non-radiative pathway is dominant and the electronic excitation is converted into rotational or vibrational motion within the polymer and its surroundings. The difference between the absorption and emission maxima of the spectra is called the Stokes shift, and it occurs when emission from the lowest vibrational excited state relaxes to various vibrational levels of the electronic ground state.

Quantitative analysis of the emission efficiency of the polymer is characterized by its quantum yield of luminescence (Φpl). The Φpl is the ratio of the number of photons emitted to the number of photons absorbed, as shown in Equation 1.1.

$$\Phi_{PL} = \frac{Photons_{EM}}{Photons_{ABS}} \qquad \textbf{(1.1)}$$

Quantum yield of photo-luminescence. Photons EM = photons emitted, Photons ABS = photons absorbed. According to the law of conservation of energy, the maximum Φpl must be 1. The value of Φpl is related to the rates of radiative (τr) and non-radiative (τnr) decays, as described in Equation 1.2. [113]

$$\Phi_{PL} = \frac{\tau_r}{\tau_r + \tau_{nr}} \qquad \textbf{(1.2)}$$

As τnr approaches 0, the quantum yield of luminescence approaches unity. Generally, Φpl is highest in dilute solutions, where the emitting species are isolated from each other. In most cases, increasing the concentration of the polymer in solution decreases the quantum yield of luminescence due to concentration quenching. Electroluminescence (EL) is emission observed upon the application of an electric field on a polymer and is observed from Light Emitting Diodes (LED). A schematic diagram of a simple LED is depicted in Figure 33. This LED consists of a transparent conducting anode (typically Indium Tin Oxide (ITO)), an emitting molecule (or polymer), and a low work function metal cathode such as calcium.

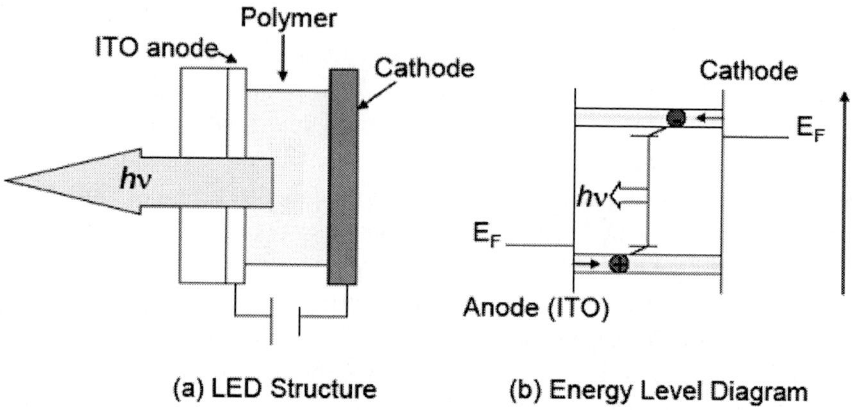

Figure 33. LED structure and the corresponding energy level diagram.

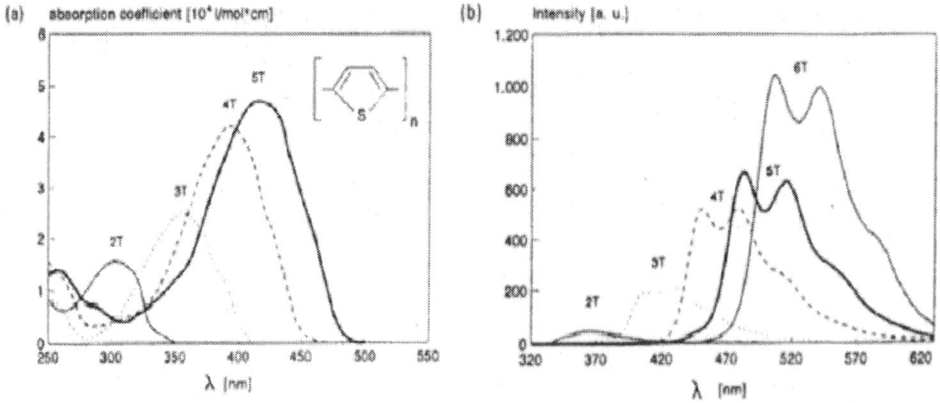

Figure 34. Absorption and emission spectra of oligothiophienes.

Light emitting diodes are bipolar devices, where holes (*h*+) and electrons (*e*-) are transported in the polymer [114]. Upon application of an electric field, a hole and electron are simultaneously injected into the polymer, whereby the hole electron pair forms an exciton. Recombination of the exciton results in the emission of light. This process is illustrated in Figure 33. Multi-layer device structures are commonly fabricated to facilitate and/or balance the hole and electron transport [115]. The mean conjugation length of the conjugated polymer is often related to the maximum wavelength of both absorption and emission spectra. This has been confirmed experimentally in acetylene, [116] thiophene, [117] phenylene [118] and phenylenevinylene oligomers [119] with various numbers of repeating units. As an example, Figure 34 depicts the absorption (a) and emission (b) spectra of oligothiophenes from 2 to 6 units long. It is clear that as the number of thiophenes increase, both the absorption and emission wavelengths increase due to larger conjugation lengths [120].

Chapter 5

NANOCOMPOSITES BASED ON CONDUCTING POLYMERS

In order to overcome the limitations for most applications of conducting polymers(CPs) like poor mechanical properties and processibility, low stability, as well as environmental compatibility and instability with respect to oxygen and/or temperature, conducting composites by mixing of the CPs with other polymers and more recent conducting nanocomposites have been developed [121-124]. The word *composite* derives from the Latin word "compositus" meaning together and "positus" meaning "put together" or "combining separate parts". Nanocomposites are the same as micro composites with one additional feature; namely, the combination of the two separate parts realized in a scale ranging from one to several hundred nanometers. The polymer nanocomposites, over the last 15 years, have been widely investigated as materials exhibiting superior properties, such as high modulus, increased thermal stability, and good gas-barrier characteristics. [125-126]. Their unique properties, with applications to automotive, mechanical, and biomedical devices, have attracted much attention from many researchers and companies.[126-131]. Nanocomposite materials hold the potential to redefine the field of traditional composite materials both in terms of performance and potential applications and have tremendous market potential both as replacements for current composites and in the creation of new markets through their outstanding properties.

Chapter 6

Classification of Nano Conducting Polymer Based Nanocomposites

Based upon the above mentioned techniques, the conducting polymer based nanocomposites can be prepared using:

6.1. Layered like inorganic media and porous hosts

6.2. Nano-scale metal nanoparticles

6.3. Nano-scale metal oxide nanoparticles

6.4. Carbon nanotubes

6.1. Layered Like Inorganic Media and Porous Hosts

One general approach to prepare polymer nanocomposites is to employ intercalation chemistry of layered inorganic solids in which polymer is inserted in to the clay galleries. Clay minerals used for polymer nanocomposites can be classified into three groups as summarized in Table.4.They are 2:1 type, 1:1 type and layered silicic acids. Their structures are briefly described in Figure 35.

The important characteristics pertinent to application of clay minerals in polymer nanocomposites are their intercalation chemistry, high strength and stiffness and high aspect ratio of individual platelets, abundance in nature and low cost. First, their unique layered structure and high intercalation capabilities allow them to be chemically modified and be compatible with polymers, which makes them particularly attractive in the development of clay-based polymer nanocomposites.

Table 4. Structures of common silicates

Type of clay	Formula	Origin	Substitution	Layer charge
2:1 type				
MMT	$M_x(Al_{2-x}Mg_x)Si_4O_{10}(OH)_2 \cdot nH_2O$	N	Octahedral	Negative
Hectorite	$M_x(Mg_{3-x}Li_x)Si_4O_{10}(OH)_2 \cdot nH_2O$	N	Octahedral	Negative
Saponite	$M_xMg_3(Si_{4-x}Al_x)O_{10}(OH)_2 \cdot nH_2O$	N	Tetrahedral	Negative
Fluorohectorite	$M_x(Mg_{3-x}Li_x)Si_4O_{10}F_2 \cdot nH_2O$	S	Octahedral	Negative
Laponite	$M_x(Mg_{3-x}Li_x)Si_4O_{10}(OH)_2 \cdot nH_2O$	S	Octahedral	Negative
Fluoromica (Somasif)	$NaMg_{2.5}Si_4O_{10}F_2$	S	Octahedral	Negative
1:1 type				
Kaolinite	$Al_2Si_2O_5(OH)_4$	N	—	Neutral
Halloysite	$Al_2Si_2O_5(OH)_4 \cdot 2H_2O$	N	—	Neutral
Layered silicic acid				
Kanemite	$Na_2Si_4O_9 \cdot 5H_2O$	N/S	Tetrahedral	Negative
Makatite	$NaHSi_2O_5 \cdot 7H_2O$	N/S	Tetrahedral	Negative
Octasilicate	$Na_2Si_8O_{17} \cdot 9H_2O$	S	Tetrahedral	Negative
Magadiite	$Na_2Si_{14}O_{29} \cdot 10H_2O$	N/S	Tetrahedral	Negative
Kenyaite	$Na_2Si_{20}O_4 \cdot 10H_2O$	S	Tetrahedral	Negative

M indicates exchangeable ions represented by monovalent ions. Symbols: N (nature) and S (synthetic).

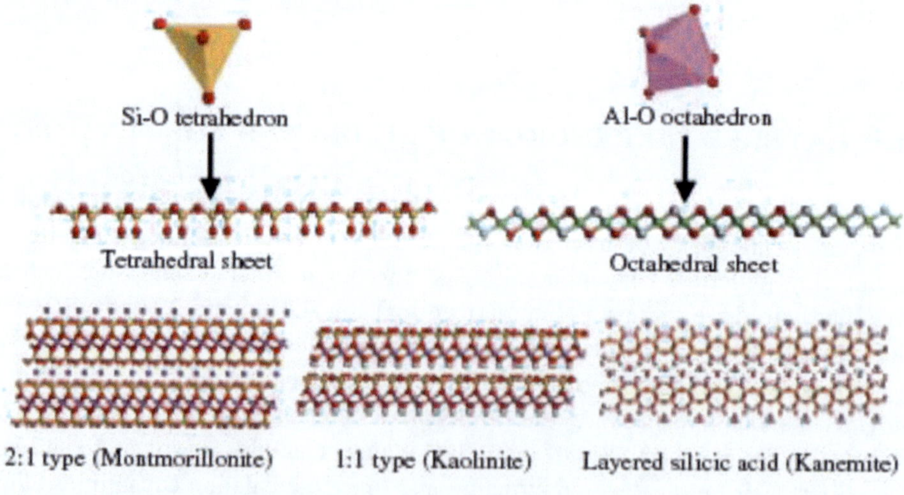

Figure 35. Structure of clay minerals represented by montmorillonite, kaolinite and kanemite.They are built up from combinations of tetrahedral and octahedral sheets ;basic units are usually Si–O tetrahedron and Al–O octahedron.

Several nanocomposites of layered silicates with conducting polymers including the N-heterocyclic polymers like polypyrrole (PPy) [132-137], poly(N-vinylcarbazole) (PNVC) [138-139], and polyaromatics such as polyaniline (PANI) [140-154] , poly(p-phenylene vinylene)(PPV) [155] and related polymers [156] have been reported. PPy and PANI are known to display electric conductivity and PNVC is well known for its high thermal stability and characteristic optoelectronic properties [157-160]. Mostly intercalation can be accomplished either by treatment with aniline via the gas phase or by immersion of the solid layered materials into liquid aniline as indicated by an increase in layer spacing from 12.6 to 15.6 Å. The first report on an attempt to polymerize the aniline by electrochemical oxidation in the intercalated materials using a clay modified electrode dates back to 1987 [161]; the reaction was reported to cause a change in colour from white to blue and a decrease in layer spacing to 13 Å. The changes were explained by the reorganization of the aniline molecules from perpendicular arrangement prior polymerization to a planar orientation of the polyaniline with respect to the silicate layers after polymerization.In the process for direct electropolymerization of PANI-MMT nanocomposite described by *Feng et al.* [148], it was confirmed that the aniline was easily oxidized between the silicate layers, and polymerized to polyaniline in the electric field; the presence of the negatively charged silicate layers stabilized the radical cation and facilitated the polymerization.*Chen and Yang* [121] prepared composites by electro-oxidizing MMT loaded with anilinium deposited on a platinum sheet electrode in the presence of further anilinium in the acidic electrolyte solution.Multilayered films of PANI were fabricated by intercalative polymerization of aniline in a synthetic mica-type silicate, a fluoro-hectorite, by the pioneer of nanocomposite materials *E.P. Giannelis* [162]. *Porter and co-workers* [163-165] described the spontaneous surface polymerization as well as intergallery polymerization of aniline while exposing layered silicate materials such as hectorite and montmorillonite to aniline vapour resulting in two-dimensional [2-D] layers of PANI formed in the gallery region The synthesis of water dispersible nanocomposites of PANI and montmorillonite was reported by *Biswas and Ray* [141] where the oxidative polymerization of aniline was performed in presence of montmorillonite dispersed in aqueous solution.However, the electrical conductivity of the nanocomposites materials (ca 10^{-2} S/cm) was found to be of two orders of magnitude lower compared to the PANI itself. The insertion of the polymer into the layers of montmorillonite clay by performing an ion exchange step (anilinium against sodium) followed by a subsequent polymerization resulted in a nanocomposite with extended chain conformation of PANI [144]. A more detailed and systematic study of the effect of the structural variation obtained by changing the PANI content during the in-situ synthesis onto the physical properties with special emphasis on the electrical properties was done by *Lee et al.* [143] using the same synthetic procedure, Figure 36, showed that the maximum expansion of the clay layers to 1.4-1.5 Å reached at a loading of PANI in the order of 12 wt %. A dramatic increase in electrical conductivity of the composite materials was observed.

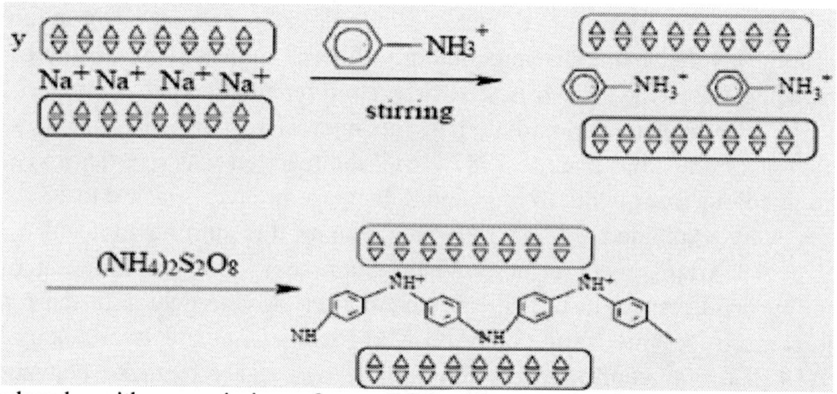

Reproduced with permission from RSC "Structural changes of polyaniline/ montmorillonite nanocomposites and their effects on physical properties, D. Lee , K. Char , S.W.Lee, Y.W. Park , J Mater Chem, 2003,13:2942-2947.

Figure 36. Intercalation and in-situ polymerization of aniline in MMT.

Different intercalation behaviours during the mechano-chemical processing suggested that the smaller the size of counter anions, the more guest molecules could be intercalated into the confined clay interlayer in highly ordered arrangements, Figure 37.The polymerization by the mechano-chemical grinding with ammonium peroxodisulfate in a mortar, decreased the basal spacing to 1.33 nm, indicating that PANI was synthesized between the interlayer spaces of MMT.

Exfoliated clay nanocomposites have also been synthesized from PANI graft copolymers by in-situ polymerization of aniline onto pre-exfoliated water-soluble poly(styrenesulfonic acid-*co*-amino styrene) (PSSA-*co*-AMS) / clay nanocomposite) or by simple blending of poly(styrene sulfonic acid-*g*-aniline) [PSSA-*g*-PANI] with clay[166]. The electrical conductivity of the nanocomposite was found to be slightly lower than that of pure PSSA-*g*-PANI, but the thermal stability and compatibility of the nanocomposite improved with PSSA-*g*-PANI, Figure 38.

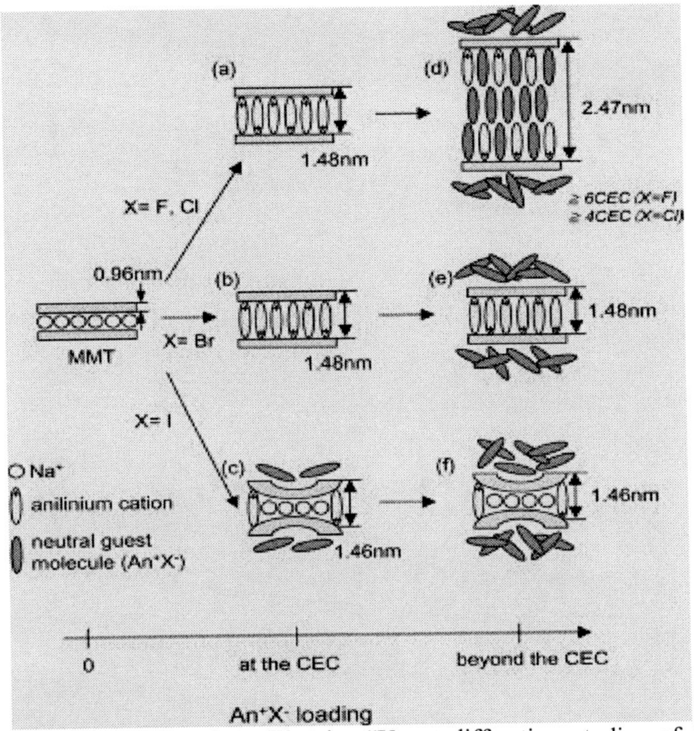

Reproduced with permission from Elsevier "X-ray diffraction studies of intercalation compounds prepared from aniline salts and montmorillonite by a mechanochemical processing" S.Yoshimoto, F.Ohashi, T.Kameyama , Solid State Commun, 2005, 136(5) , 251-256.

Figure 37. Stacking of the intercalated species in the clay layers.

The synthesis of polypyrrole-montmorillonite clay composite by chemical method in aqueous medium has also been investigated. MMT clay-based composites of PPY were successfully obtained by polymerizing pyrrole in the presence of $FeCl_3$-impregnated MMT clay in aqueous medium.

The synthesis of PANI-MMT nanocomposite using different dopants has always been the subject of intense research as various morphological structures were obtained upon altering the dopant, Figure 39.

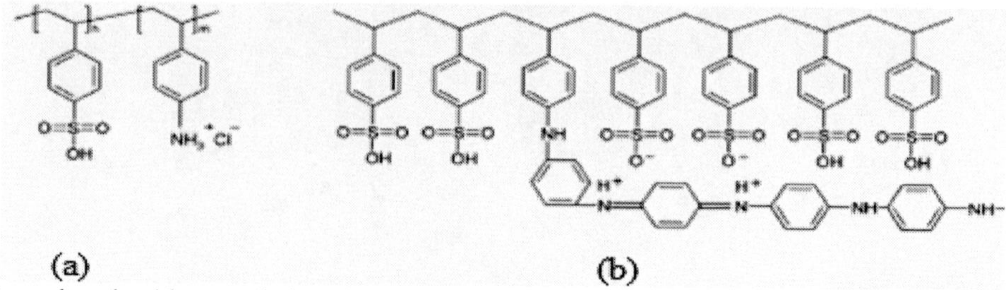

Reproduced with permission from ACS "Exfoliated Nanocomposite from Polyaniline Graft Copolymer/Clay" W.J. Bae, K.H .Kim, W.H. Jo, Y.H. Park Macromolecules, 2004, 37(26), 9850-9855.

Figure 38. Chemical structures for (a) PSSA-co-AMS, and (c) PSSA-g-PANI.

Figure 39. SEM micrographs of PANI/MMT doped with different dopants (a) PANI-DBSA, b- PANI-NSA, c- PANI-PTSA, d- PANI-SA, e- PANI-CSA, f- PANICN-NSA, g-PANICNPTSA, PANICNSA, i- PANICN-CSA, j- PANICN-DBSA.

The conductivities of PPY/MMT composites increased with the increase in MMT clay loading in the composite [167]. The morphological study from the SEM analysis, Figure 40 shows the denser and more compact morphology of PPy/MMT clay composite with the increase in MMT clay which was found to be responsible for enhancing the conductivity [167].

Figure 40. SEM micrograph of PPY/MMT clay (a) (0.4 % clay).

6.1.1. PANI-silica Nanocomposites

Highly homogeneous and transparent glass-PANI nanocomposites were obtained by polymerization of aniline, adsorbed in its protonated (i.e. positively charged) form via an ion-exchange process within the pores of porous vycor glass by the use of ammonium persulfate acidic solution as oxidant [168]; the use of Cu^{2+} ions absorbed inside the glass pores yielded only oligomeric PANI [169]. As evidenced by ^{29}Si NMR spectrometry, SiO_2 groups of the glass acted as counter ions for the doping of PANI. The glass frame was removed by treatment with HF and the resulting PANI morphology was formed only inside the pores, Figure 41.

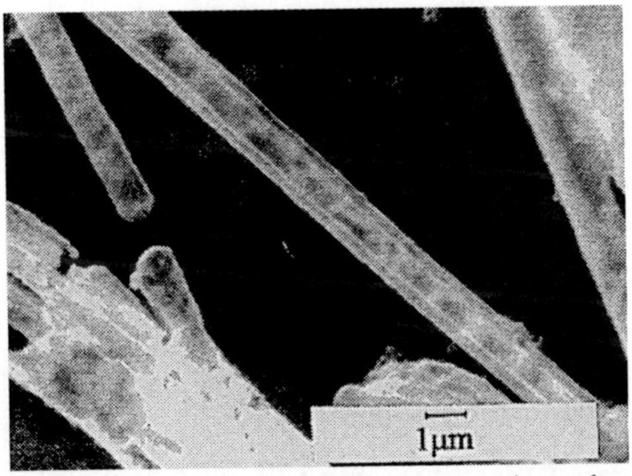

Reproduced with permission from Elsevier "Nanocomposites glass/conductive polymers "A.J.G. Zarbin, M-A De Paoli, O.L. Alves Synth. Met., 1999, 99(3), 227-235.

Figure 41. SEM micrograph of PANI extracted from PANi / Porous Vycor Glass nanocomposite.

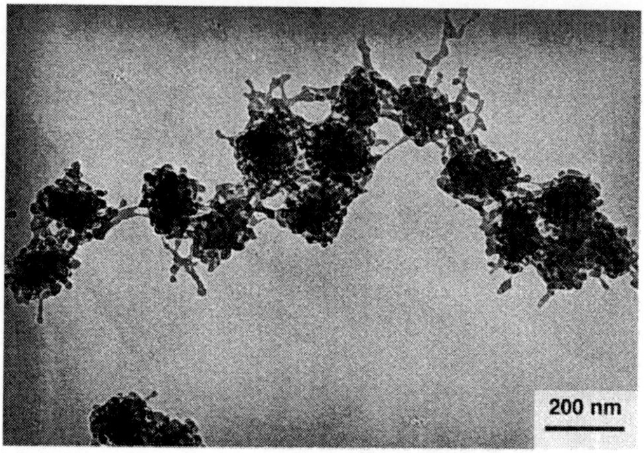

Reproduced with permission form Elsevier "Synthesis of Colloidal Dispersions of Polypyrrole-Silica Nanocomposites Using "Stringy" Silica Particles" R. Flitton, J.Johal, S.Maeda, S.P.Armes J Colloid Interf Sci,1995, 173,135-143.

Figure 42. TEM of diluted PANI – silica colloid, synthesized using the $(NH_4)_2S_2O_8$ oxidant (silica concentration of 2.0%).

The material formed was highly organized PANI filaments with diameters much larger than the pore sizes (ca. 83 Å) and the intrinsic filament diameter of ca 25 Å indicating an aggregation tendency of the free polymer wires. Numerous reports have described the deposition of PANI or their derivatives on micro- and nanometer-sized silica particles. *Armes and co-workers* reported on the preparation and characterization of polyaniline-silica particles using about 20 nm spherical silica particles as particulate dispersants in aqueous media [170-174]. These ultrafine oxide sols act as high surface area colloidal substrates for the precipitating PANI, resulting in stable colloidal dispersions of raspberry-shaped particles in the 100-300 nm size range. Using non-spherical 'stringy' silica particles, stable colloidal dispersions were formed , but the morphology was rather more ill-defined than that of the raspberry nanocomposites obtained using spherical silica particles.The colloidal 'rasberries' consist of micro-aggregates of silica 'glued' together by the conducting PANI component, Figure 42. X-ray photoelectron spectroscopy showed that the surface of the conducting polymer-inorganic oxide hybrid particles are oxide-rich, thus their excellent long-term colloidal stability will be due to a charge stabilization mechanism. Small-angle X-ray scattering revealed that the average silica-silica separation distance within the raspberry particles is about 4 nm, suggesting that the PANI chains are dispersed at the molecular level, making suchparticles true nanocomposites. The conductivity, measured on pressed pellets, was about 3 $S.cm^{-1}$. BET surface studies show that the specific surface areas of the composites are substantially higher than estimated from TEM studies, even allowing for the unusual 'raspberry' morphology. Therefore, these nanocomposite systems were confirmed to have significant porosity. Following the procedures of *Maeda and Armes* [173], *Neah et al.* [175] prepared PANI-SiO_2 nanocomposites for the uptake of gold and palladium from $AuCl_3$ and $PdCl_2$ in acid solution respectively. In case of gold uptake, the reaction rate increased with temperature from 0 to 60°C. The accumulation of elemental gold on the nanocomposite increased the diameter and decreased the surface area. High rates of palladium uptake can only be achieved with PANI reduced to its lowest oxidation state, and unlike the uptake of gold, the palladium on the micro particles exists in the Pd(II) state.*Natase et al.* [176] deposited PANI-SiO_2 nanocomposites through a plasma polymerization route. The films were generated by spraying of mixtures of 10 nm sized silica particles dispersed in freshly distilled aniline monomer into a plasma stream of the DC-plasma discharge reactor. AFM revealed a homogeneous morphology with a very fine distribution of nanoparticles, Figure 43, coated with PANI in its emeraldine base form.The silica in the resulting polymer matrix resulted in a ballistic conduction mechanism combined with a continuous transition from shallow traps to deep traps ending with saturation.Nano PANI-SiO_2 composite was prepared by in-situ polymerization [177]. Nanosized SiO_2 particles were ultrasonicated for 2 hours in a solution of aniline in 2 M HCl.

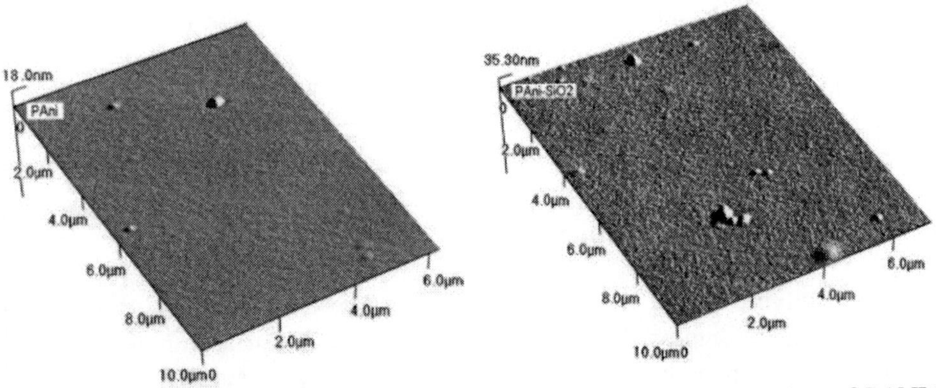

Reproduced with permission form Elsevier "Synthesis and characterization of PANI" F.Nastase, I.Stamatin, C.Nastase, D.Mihaiescu, A .Moldovan, Prog .Solid. State.Chem., 2006, 34(2-4), 191-199.

Figure 43. AFM images of PANI (left) and of PANI-SiO_2 film (right) made by plasma polymerization.

$(NH_4)_2S_2O_8$ was added to the solution and the polymerization allowed to proceed for 3 h at room temperature. PANI was found to be deposited on the surface of the nano-SiO_2 particles, forming a core-shell structure. The dimensions of the nanoparticles were about 20-30 nm, and the conductivity of particles containing 15 wt % PANI reached 0.3 $S.cm^{-1}$. Spectroanalyses revealed an interaction at the interface of SiO_2 particles and PANI macromolecules. Aniline was chemically polymerized in the presence of ultrafine colloidal silica particles with a diameter of 35 nm, and colloidally stable PANI / silica particles were obtained [178].Besides PANI, poly(o-anisidine) coated silica particles were obtained by chemical polymerization using ammonium persulfate in the presence of silica particles [179]. A faster polymerization reaction of aniline using sodium persulfate and potassium iodate as oxidants was reported to proceed in the presence of ultrafine 20 nm silica particles to form PANI / silica colloidal composites, as compared to the same reaction performed without the addition of silica [180]. Both the isoelectric point and zeta potential of PANI/silica, prepared by chemical polymerization in the presence of commercially available silica particles, were found to be governed primarily by the nature of the charged groups of silica surface rather than by the CEP component [172]. This suggested that silica is a major component at the particle surface, which was consistent with the good long-term colloidal stability of these composites.PANI grafted silica nanoparticles were prepared by the in-situ oxidation with ammonium peroxodisulfate on the surfaces of the aminopropyl silica nanoparticles ultrasonicated for 30 min [181]. After a polymerization period of 4 hours while stirring, a percentage of grafting of 205.2% and a grafting efficiency of 37.7% were achieved, calculated from the results of elemental analysis. FTIR and UV/Vis confirmed the chemical grafting of PANI.

6.1.2. PANI-silane Nanoparticles

Graft polymerization was used to prepare a dense conductive PANI coating on free standing luminescent silicon nanoparticles [182]. The silicon nanoparticles maintained their photoluminescence and crystallinity after surface modification. The nanoparticles were first surface hydroxylated and then reacted with (3-bromopropyl) trichlorosilane to form a dense bromopropylsilane monolayer. This was further reacted with aniline, which displaced the bromine atoms. The surface-bound aniline molecules acted then as active sites for the graft polymerization of PANI, Figure 44.

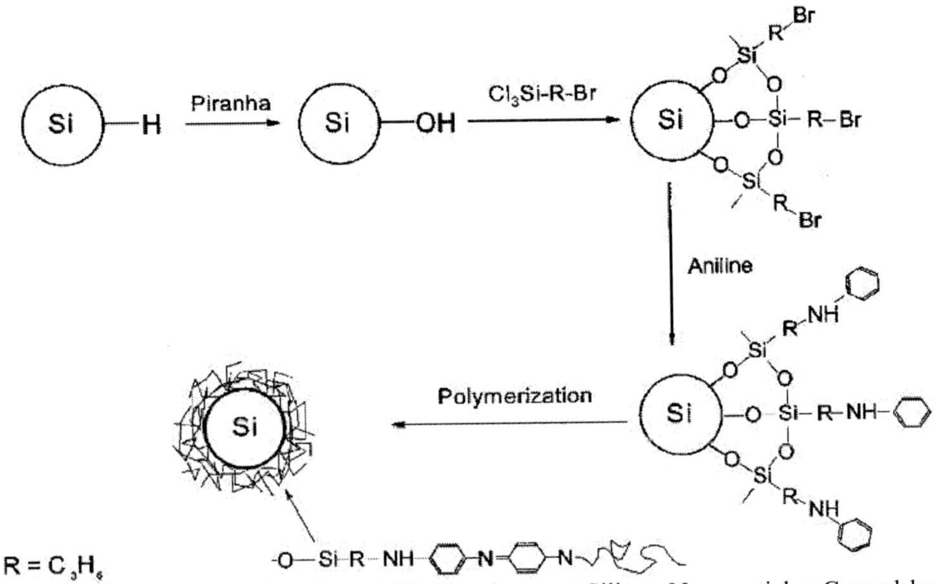

Reproduced with permission form ACS "Luminescent Silicon Nanoparticles Capped by Conductive Polyaniline through the Self-Assembly Method" Z.F. Li, M.T. Swihart, E. Ruckenstein, Langmuir, 2004,20,1963-1965.

Figure 44. Grafting of PANI on Si nanoparticles through self-assembly method, consisting of formation of Si-OH groups.

The silane self-assembled monolayer effectively protected the silicon particles against photoluminescence quenching and degradation in basic solutions that rapidly quench the photoluminescence of unprotected particles. The PANI coating enhanced this protection, even when in its nonconducting emeraldine base state. The electrical conductivity of HCl-doped (emeraldine salt) PANI-capped silicon nanocomposite exceeded 10^{-2} Scm^{-1}, which is 6 orders of magnitude higher than that of bare silicon particles. The photoluminescence spectrum and lifetime were not affected by the conductive PANI coating, indicating that charge carriers ware not transported from the silicon core to the PANI shell [182].

6.2. Nano-scale Metal Particles

Nanocomposites based on conducting polymers such as polyaniline and metal can be divided into two types. The composites of the first type are prepared by polymerization of the appropriate monomers in the presence of oxide dispersion, where the latter appear embedded in matrices of conductive polymers, which serve simultaneously as conductor and binder. The composites of the second type are guest-host compounds, when macromolecules of conductive polymers are inserted into the interlayer galleries or channel of inorganic particles. Three methods of preparing such nanocomposites are known.

- Intercalation of monomers into the host matrix followed by polymerization due to an external effect (e.g. interaction with an oxidant).
- Redox based host intercalation of the monomers and in-situ polymerization.
- Direct intercalation of polymer macromolecules into the host particles.

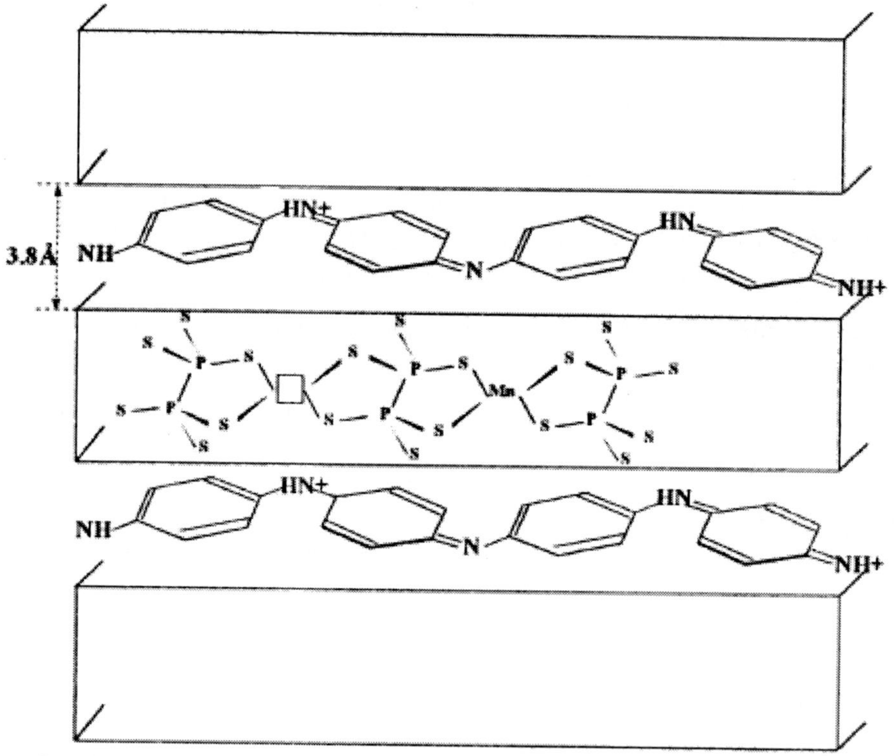

Reproduced with permission from Elsevier "Preparation of a new nanocomposite of conducting polyaniline into layered MnPS3 " D. Zhang, J .Qin, K .Yakushi, Y. Najazawa, Mater Sci Eng A, 2000,286,183-187.

Figure 45. PANI-MnP_3 intercalates.

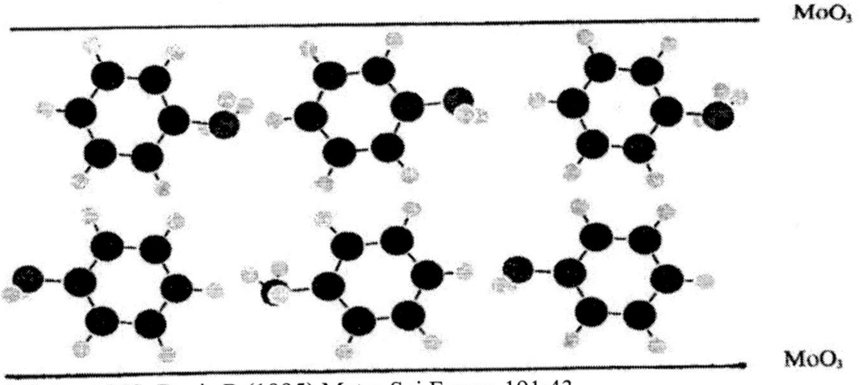

Hill PG, Foot PJS, Davis R (1995) Mater Sci Forum 191:43.

Figure 46. Arrangement of the aniline bilayer in MoO_3.

A new family of intercalates with host layers of MPS_3 (M = Mn, Cd) and guest layers of PANI have been prepared by ion-exchange process [183-184]. MPS_3, first expanded by K^+ ions, reacted with PANI-NMP solutions, yielding intercalate with basal spacing of 9.3 Å [183] and 10.3 Å [184], Figure 45. Nanocomposites $(PANI{\cdot}CSA_{0.124})_{1.06}MoO_3$ have also been investigated by adding a *m*-cresol solution of PANI·CSA to an aqueous sol of Li_xMo_3 during mechanical stirring for 1 hour at room temperature [185]. The data analysis showed a complete absence of lithium in the composite and the amount of *m*-cresol and water equal to 0.42 and 9.5 wt %, respectively. The nanocomposite has a conductivity of 6.4 x 10^{-1} S/cm.A nanocomposite of PANI with MoO_3 produced by polymerization of aniline in the presence of a suspension of MoO_3 in an aqueous acidic medium (H_2SO_4) using ammonium vanadate ([AV], V^{5+}) as the oxidant was described by *Ballav and Biswas* [186]. The oxidation of an aniline-MoO_3 intercalate was studied by *Hill et al.* [187] , Figure 46, using aqueous H_2O_2, saturated iodine vapour and dried air. The intercalate was made by ion exchange of hydrated sodium ions.

Conducting PANI / tungsten oxide composites have also been synthesized by in-situ polymerization [188]. Ammonium persulfate was used as the oxidant in the mixture of aniline and fine graded WO_3. Aniline was found to intercalates readily in layered Bronsted acids, $HMMoO_6{\cdot}H_2O$ (M = Nb, Ta) possessing a rutile-type structure [189]. Hybrid materials such as polyoxometallate $H_3PMo_{12}O_{40}$ were also prepared from and aniline. *Kulesza et al.* [190] described the preparation of ultra thin PANI mono- and multilayer on an electrode surface by repeated and alternate treatments with solutions of dodeca molybdo phosphate anions and anilinium cations, and subsequent electrochemical polymerization.Nanocomposites based on

$NbWO_6$ nanosheets and aniline were synthesized by an exfoliation-reflocculation method [191]. Rutile-type $HNbMoO_6$ was intercalated with aniline, and subsequent polymerization was performed using $FeCl_3$, resulting in a composite material $(PANI)_yNbMoO_6$, where y ranges from 0.26 to 0.64 [192]. Alternatively, tri rutile-like layered $HMWO_6{\cdot}nH_2O$ (M = Ta, Nb) was used for intercalation with aniline, forming a bilayer of the guest species within the interlayer gap [193].These nanocomposites have been used as cathodes and improved the electrochemical lithium insertion in conventional lithium cells. Layered perovskites, $HCa_2Nb_3O_{10}$ and $HCa2Nb_2M_9$ (M = Al, Fe), are able to intercalate aniline and anilinium through an acid-base interaction [194]. A subsequent polymerization by treatment of the intercalate with $(NH_4)_2S_2O_8$ gives novel PANI / layered perovskites composites. The polymer exists within the galleries of the host in the emeraldine salt form with considerable cross-linking. The incorporation of the conductive emeraldine form of PANI in the interlayer distance of tin(IV)-2-carboxyethylphosphonate by exposure of the Fe(III) intercalated host to an aqueous solution of aniline results in a composite showed the characteristic electro active properties of PANI [195]. Two approaches were reported for the preparation of PANI-containing nanocomposite with γ-$Sn(HPO_4)_2 \bullet 2\,H_2O$ [196]. First, the inorganic material was charged with Fe^{3+} ions, and then used in oxidative polymerization of aniline, proceeding inside the layers of inorganic material, or, secondly, aniline was intercalated in the material, and polymerization was performed by an ammonium persulfate solution. The second procedure resulted in higher contents of intercalated PANI (10.8 wt % compared to 2.2 wt %), however the amount of polyaniline formed is by far smaller than the amount of aniline intercalated prior to the polymerization reaction (18 %). The interlayer distance increased by 6.6 Å after aniline intercalation and returned to its original value after polymerization. Cyclic voltammetry experiments hint to the existence of highly linear chains with negligible crosslinking in the nanocomposite materials. The α-$RuCl_3$ was successfully used as a host matrix for the in-situ redox intercalative polymerization of aniline resulting in a nanocomposite material with a positive Seebeck effect suggesting that holes were the dominant carriers in the electrical transport [197].

6.2.1. PANI-Au an PANI-Ag Nanoparticles

Traditionally, the synthesis of metal nanoparticles was carried out chemically with a reducing agent (such as $NABH_4$) or with irradiation to convert groups of metal ions into nanoparticles. However the processes suffered from low yield and low stability in the air-oxygen environment. Therefore, other synthesis routes were explored. Synthesis of Ag nanoparticles was achieved by an incremental addition of a metal ion aqueous solution to the PANI colloid solution while stirring. When using undoped PANI, the Ag particles were dispersed homogeneously in the solution with some particles residing at the periphery of the colloids, Figure 47.When using nitric acid doped PANI, the synthesized Ag nanoparticles were located strictly within the core of the colloids. The size of 75% of the Ag particles ranges from 2.5 – 4.5 nm if $AgNO_3$ was added incrementally and the resulting solution stirred. Much larger nanoparticles were formed if the concentration of $AgNO_3$ was increased, and even nanosheets up to 250 nm with a thickness of ~20 nm could be observed if all $AgNO_3$ was added at one time.

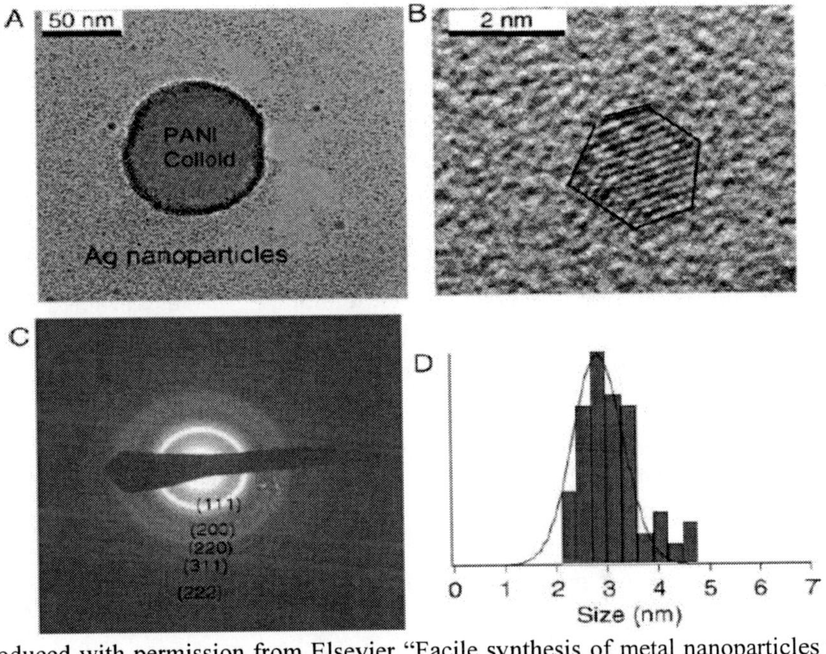

Reproduced with permission from Elsevier "Facile synthesis of metal nanoparticles using conducting polymer colloids", W Li, QX Jia, H-L Wang, Polymer 2006,47,23-27.

Figure 47. (A) TEM image showing an undoped PANI colloid and homogenously dispersed Ag nanoparticles. (B) Magnification of Ag particles. (C) Graph showing the electron diffraction of an ensemble of Ag nanoparticles. (D) Size distribution histogram of the Ag nanoparticles.

Del Castillo-Castro and colleagues [198] synthesized Ag and Cu-Ag nanoparticles by reducing tetrachloroauric acid with hydrazine in water-in-oil micro-emulsions of water and HAuCl4-isooctane; for the synthesis of Ag particles sodium bis(2 ethyl hexyl) sulfosuccinate (Aerosol OT) was added as surfactant, and for Cu-Ag particles a mixture of copper(II) bis(2-ethylhexyl) sulfosuccinate $[Cu(AOT)_2]$ and AOT. The incorporation of Ag particles increased the electrical conductivity by two orders of magnitude and for Cu-Ag particles by one order of magnitude. The same authors also reported the synthesis of Au nanoparticles in presence of undoped and doped PANI. In this case, the nanoparticles were aggregated at the outer surface of the undoped PANI colloids or trapped inside the doped PANI colloids. The average size of the Au nanoparticles was estimated to be 7-12 nm. Furthermore, the synthesis of 2 nm Pt nanoparticles was demonstrated. CTAB-stabilized reverse micro-emulsions were used in different studies to synthesize Au colloids [199-200].*Lin et al.* [199] used a CTAB/*n*-hexanol/water system and produced single Au crystals, however, of many different shapes (cylinder, triangle, sphere, hexagon, trapezoid, and pentagon) with a broad size distribution (7 – 40 nm), due to aggregation and growth of bare gold particles. Self-organization of the Au particles could only be observed when dodecanethiol-extraction and redispersion in toluene were applied, but also a large amount of black precipitate containing Au resulted. Using quaternary CTAB-stabilized reverse micro-emulsions, i.e. CTAB/pentanol/hexane/ water, *Chen et al.* [200] were able to synthesize monodisperse colloidal Au nanoparticles (about 7 nm) in hexane. Self-assembled linear/circular shapes and PANI-induced hexagonal patterns were observed on carbon-coated copper grids. Compared with the results of those obtained in CTAB/*n*-butanol/octane/water reverse micro-emulsions [199] , the homogeneity in size and shape of the Au nanoparticles was attributed to the strong protection of a compact CTAB/*n*-pentanol interfacial film. Au nanocrystals dispersed in organic solvents, particularly in hexane, have wide ranging applications, such as catalysts for the growth of silicon, or GAP nanowires, templates for the formation of hollow polymers, and building blocks for the fabrication of superstructures. The electrical response of PANI-PBMA-Au nanoparticles films to peroxide and ammonia exposure suggests applications as uric acid and urea biosensor.Nanostructured composite materials of PANI with gold clusters of different sizes and amounts were also produced by *Smith et al.* [201]. PANI films of identical thickness were first grown electrochemically on platinum substrates. Next, nucleation of gold clusters was initiated by chemical reduction of tetrabromoaurate ion in acidic solution. It was found that the amount and the size distribution of spontaneously formed gold clusters varied with immersion time. The morphology of these films consisted of a dense network of

PANI fibrils or bundles with small Au spherulites attach along their surface. The cluster surfaces are largely open to solution and may be further functionalized. Despite the fact that the preparation method makes gold cluster growth difficult to control in size and distribution, the nanocomposites are interesting for applications such as batteries and super capacitors, or catalysis.*Zou et al.* [202] synthesized mercaptoethane sulfonate protected gold and silver nanoparticles (Au-MES, Ag-MES) by the modified procedure of *Brust et al.* [203] and Chen and Kimura [204] Both Ag-MES and Ag-MES nanoparticles are soluble in water up to 2.0 mg/ml, and the stability of Au-MES is much better than that of AG-MES. The emeraldine base of PANI, prepared by the method proposed by *Cheung et al.* [205] was dissolved in dimethylformamide [DMF] solution (~2.0 mg/ml). Quartz slides were first dipped into this PANI in DMF solution, and, after rinsing and drying, subsequently in a 1.0 mg/ml of Au-MES aqueous solution containing 0.01 M HCl, and a polyanion monolayer (i.e. Au-MES nanoparticles on top) was formed. This process was repeated several times to form multilayers of (Au-MES)-PANI. (Ag-MES)-PANI multilayers could be fabricated in a similar way. Highly conducting PANI films can be obtained by doping with these nanoparticles. Ag-PANI nanocomposites were prepared by *Kang et al.* [206] via two different methods, both using γ-ray irradiation. In method 1, polyvinylpyrrolydinone [PVP] – stabilized Ag colloids were prepared by γ -irradiation; aniline was added and polymerized oxidatively. In method 2, aniline-stabilized Ag colloids were prepared by γ-irradiation followed by oxidative polymerization. A large numbers of hydrated electrons produced by γ -ray irradiation can reduce the metal ions to zero valent metal particles, avoiding the use of additional reducing agent and the consequent side reactions. The amount of zero-valent nuclei can be controlled by varying the irradiation dose. The formation of many nuclei is favourable to result in highly dispersed nanoparticles. Method 1 resulted in sphere-type nanoparticles with an average size of 13 nm. The morphology of the nanoparticles made by method 2 was found to change from hexangular-type particles to sphere-type particles when increasing the amount of aniline to stabilize the Ag nanoparticles. Manipulation of the experimental conditions opens up the scope for fine tuning the morphology of Ag-PANI nanocomposites.*Pillalamarri et al.* [207] used γ -ray irradiation to carry out a one-pot synthesis of PANI nanofibres decorated with noble-metal (Ag or Au) nanoparticles. Aqueous solutions of aniline and the corresponding metal salt were irradiated; radiation doses levels of 3.5 kGy, at an initial dose rate of 1.5 kGy/h, were used. Exposure to higher doses did not alter the morphology. The aniline polymerized as very thin fibres (typical diameters of 50 – 100 nm and lengths of 1 – 3 nm); the shape and size of the metal particles could be changed from nanometer-sized spheres to micron-sized dendrites by varying the ratio of aniline to the metal precursor. The electrical conductivity of the composites increased with the loading of metal in the nanocomposites and was for Ag particles up to 10 times greater and for Au particles up to 50 times greater than that of PANI fibres alone. *Khanna et al.* [208] used UV-radiation, Figure 48, in the preparation of Ag-PANI nanocomposites.An in- situ reduction of silver salt in aniline was carried by mild photolysis performed with long wavelength (365 nm) and short wavelength (254 nm) radiation. The reduction of silver nitrate salt in aqueous aniline leads to the formation of nanoparticles of silver that later act as photo catalyst to oxidize aniline to polyaniline. Under the UV light source of 254 nm wavelength the reaction turned out to be faster than under the visible light source of 365 nm wavelength. Small particles (size ~30 nm) of cubic silver were indicated in XRD. SEM revealed a uniform size distribution with spherical and granular morphology and TGA showed the improved thermal stability of the PANI nanocomposite.

Figure 48. SEM of Ag-PANI nanocomposite prepared by photolysis method at 254 nm; (a) 1 M $AgNO_3$, (b) 2 M $AgNO_3$. (c) SEM image of Ag-PANI nanocomposite prepared at 365 nm (1 M $AgNO_3$).

Zhou et al. [209] used the unsymmetrical square wave current method [USCM] to synthesize Ag-PANI nanocomposite films. USCM is characterized by the combination of an anodic process of aniline monomer polymerization and a cathodic process of metal ionic electro-deposition. Ag-PANI nanocomposite films prepared by USCM had fibrous structure and the Ag nanoparticles were not only dispersed on the surface of the PANI film but also embedded in the PANI matrix. The nanocomposite film showed considerably higher electro activity – lower charge transfer resistance, higher redox peak current – compared with PANI film without Ag particles. The electrochemical property of Ag-PANI nanocomposite films was manipulated with synthesis parameters, such as cathodic current density and total synthesis time. A sonochemical method was used by *Sivakumar and Gedanken* [210] to synthesize polyaniline and Au-PANI composites. However, no formation of PANI was observed without the external addition of H_2O_2. The addition of peroxide led to a significant reduction of the reaction time (1/48 times) and a significant increase in yield (~20 times). The added H_2O_2 also resulted in an increased generation of hydroxyl radicals by ultrasound, which facilitates polymerization. Ultrasound results also in a more homogenous size distribution in both PANI and the Au-PANI composite (particles for both about 30 nm) when comparing the preparation method with conventional stirring. Xian et al. [211] described a glucose sensor based on a nanocomposite consisting of Au nanoparticles and conductive PANI nanofibres. Because of the large specific surface area, excellent conductivity and the presence of many micro-gaps in the nanocomposite,GOx could be immobilized with high loading and activity.Nafion was used as a coating for eliminating the interferences of electro active compounds to the glucose response. The glucose sensor shows a linear calibration curve, with a low detection limit and rapid response time. Moreover, the nanocomposite sensor exhibits high stability and excellent reproducibility [212].

6.2.3. PANI-Pt Nanoparticles

Zheng and Li [213] developed a new route to fabricate Pt-PANI nanofilms at a liquid/liquid interface in a reverse micelle using H_2PtCl_6 and aniline. Firstly H_2PtCl_6 was incorporated in AOT reverse micelles reduced to Pt nanoparticles by the protonated aniline, while the protonated aniline was oxidized to anilinium radical cation. The Pt nanoparticles served as the catalyst for the further polymerization of aniline. Therefore, the number of steps in the complex set of sequential reactions leading to the formation of the composite was reduced.

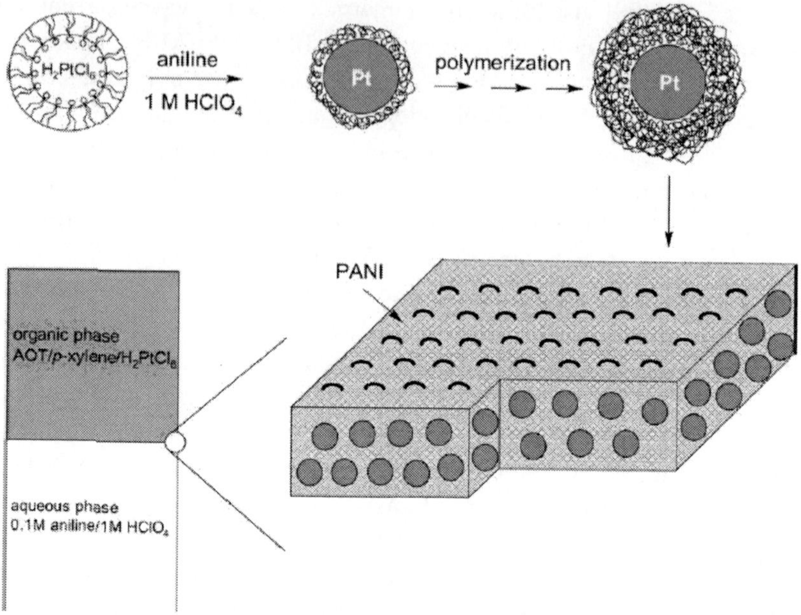

Reproduced with permission from Elsevier "Platinum–polyaniline nanofilms synthesized at a liquidliquid interface with enhanced conductivity", L.Zheng, J.Li ,J Electroanal Chem 2005,577,137-142.

Figure 49. Mechanism of formation of Pt-PANI nanofilms.

Furthermore, a liquid/liquid interface was chosen for the site of the formation of Pt-PANI nanofilms, because such an interface offers a unique microenvironment for confining the 2-D growth of PANI, Figure 49. The nanofilms showed enhanced electrical conductivity compared to the chemically polymerized PANI. The conductivity of Pt-PANI nanofilms was found to be potential-dependent and provided promising reversible potential-switchable property.

6.2.4. PANI-Pd Nanoparticles

Huang et al. [214] showed that their method to prepare PANI (and PPy) containing gold in elemental form from acid solution in an electro less way [215-216] which was also used to synthesize PANI (and PPy) containing either Pd(0) or Pd(II) ions. They found that the rate of Pd uptake was significantly higher when the polymers were reduced to their lowest oxidation state with acid solutions containing Pd(II) ions. The choice of anions determined the reaction rate and the state of the Pd incorporated into the polymer matrix. The Pd-PANI samples had a higher specific surface area compared to commercially available aluminium oxide supported Pd and 'pure' $PdCl_2$. In the catalytic hydrogenation of nitrobenzene the reaction rate and the aniline yield in the presence of the Pd-PANI samples are comparable to those achievable with the aluminium oxide supported Pd and are significantly higher that those with $PdCl_2$. The state of Pd and the PANI support did not change significantly after the hydrogenation of nitrobenzene, demonstrating the potential use of these systems in electro-catalysis. Houdayer et al. [217] used a novel, one-pot route to prepare PANI-supported Pd(0) nanoparticles. Nanosized Pd(0) particles were first prepared by reduction of Pd(OAc)2 using t-BuONa activated sodium hydride in refluxing THF.A ligand exchange with aniline on t-BuONa-stabilized Pd(0) particles yielded aniline-stabilized particles. PANI-Pd(0) nanocomposites were finally obtained by polymerizing aniline-stabilized Pd(0) particles using ammonium persulfate.Preliminary results indicated that Pd(0) particles supported on PANI behave as efficient heterogeneous catalysts in the Heck and Suzuki-Miyaura reactions of aryl iodides. Athawale et al. [218] showed that a Pd-PANI nanocomposite was sensitive for methanol vapours and could be used as a selective methanol sensor. Pd nanoparticles were obtained by a reflux method, and the Pd-PANI nanocomposite was synthesized by oxidative polymerization of an aniline solution containing the Pd particles. The electric resistance response of the sensor changed rapidly and reversibly in the presence of methanol vapours. An explanation for this sensitivity was found in the larger surface area provided by the Pd nanoparticles in the nanocomposite near the imine nitrogens.

6.2.5. PANI-Ni and PANI-Sn Nanoparticles

Houdayer et al. [219-220] synthesized also Ni (0) and Sb(0)-PANI nanocomposites involving the reduction of $Ni(OAc)_2$ and of $SbCl_5$ or $SbCl_3$, respectively. Ni(0) particles of ca. 6 nm were found to be homogeneously distributed in PANI, Figure 50.

The catalytic activity was evaluated in the Heck reactions, showing a much higher rate of coupling between aryl diodes and olefins compared to the rate obtained with Ni catalysts. The Sb (0) precursor, was found to be useful for the electronic applications because of its highly anisotropic behaviour, low conduction band and high electron mobility, which had a great effect on size and crystallinity of the PANI-stabilized Sb(0) particles.

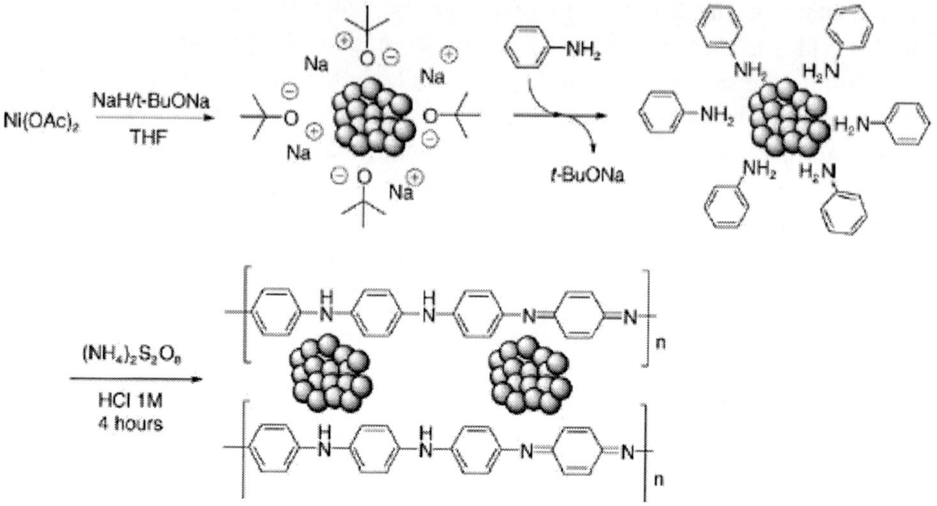

Reproduced with permission from Elsevier "New polyaniline/Ni(0) nanocomposites: Synthesis, characterization and evaluation of their catalytic activity in Heck couplings", A.Houdayer, R.Schneider, D.Billaud, J.Ghanbaja, J.Lambert, Synth Met, 2005, 151,165-175.

Figure 50. One pot' synthesis of Ni(0)-PANI nanocomposites.

Amorphous Sb (0) particles with an average diameter of 2.0 nm were obtained with $SbCl_5$, while rhombohedral particles with an average diameter of about 9.0 nm resulted using $SbCl_3$.

6.2.6. PANI-Cu Nanocomposites

Athawale et al. [221] synthesized copper nanoclusters and Cu-PANI nanocomposites. Sharma et al. [222] found chemically synthesized Cu-PANI nanocomposite to be selective for chloroform vapours. FT-IR spectra showed remarkable modifications in the far IR region indicating the interaction of chloroform with the metal cluster Cu nanoclusters were fabricated by reduction of an aqueous copper salt solution by sodium borohydride. The Cu-PANI nanocomposite was prepared by polymerizing a monomer aniline hydrochloride solution containing the Cu nanoclusters using ammonium persulfate as the oxidizing agent. The Cu clusters were found to be ~53 nm in size. The nanocomposite has been used as a catalyst in the conversion of 1-decene to 2-decanone in a molecular oxygen atmosphere, although the yield was poor. Still, it was demonstrated that such nanocomposites might be used in industrially important conversions of alkene to ketone or aldehyde.

6.3. Metal Oxides Based Nanocomposites

6.3.1. PANI-V_2O_5 Nanocomposites

Conducting PANI-V_2O_5 nanocomposites have been prepared by in-situ intercalation and polymerization of aniline or its salt in V_2O_5 xerogel by *Kanatzidis, Wu and co-workers* [223-224]. A monolayer of PANI was concluded to be present in the V_2O_5 framework based on the interlayer separation of 0.56 nm [224]. The crucial role of oxygen in this process was pointed out, and it was shown that oxygen acted as an electron acceptor both during the in-situ reaction, and after intercalation. The conductivity of the freshly prepared composite was in the range 10^{-4}-10^{-1} S/cm depending on the degree of polymerization. [224].Wu et al. [225-227] studied nanocomposites of poly(aniline-*co-N*-(4-sulfiphenyl)aniline) [PAPSA] and V_2O_5. An aqueous solution of the water soluble PAPSA was mixed with V_2O_5 wet gel, the mixture centrifuged and the $(PAPSA)_xV_20_5$ solution cast on glass or silicon substrates. After water evaporation a flexible conducting composite film resulted. Two types of films with different solubility were identified with XRD: a water-insoluble film (*d*-spacing of 11.5±0.5 Å) and a water-soluble film (*d*-spacing of 13.5±0.5 Å). The $(PANI)_{0.5}V_2O_5\bullet 1.0H_2O$ nanocomposite prepared was found to be locally well ordered and described in terms of a small number of structure-sensible parameters Figure 51.

Both ion-exchange and intercalated sites were reported for PSPAN/ V_2O_5 nanocomposite The results showed that the notions of self-doping for conducting polymers could be equally well applied to nanocomposites. Nanocomposites of conducting PANI and V_2O_5 were also reported by intercalation of the conducting polymer into the layers of V_2O_5 xerogel by *De et al.* [228] to study the charge transport mechanism. PANI of molecular size 5 Å was confirmed within the interlamellar space of the gels. A manipulation at the molecular level of PANI-V_2O_5 nanocomposites built by the layer-by layer technique promoted the enhancement of charge storage capacity, new electrochromic effects, and control of ionic flux as described by *Huguenin et al.* [229-230].Pang et al. [231] synthesized PANI-V_2O_5 nanocomposite nanosheets with thickness between 10 and 20 nm, and lateral dimensions in the range of hundreds of nanometers to several microns by in-situ intercalation polymerization of aniline with layered V_2O_5 under hydrothermal conditions.

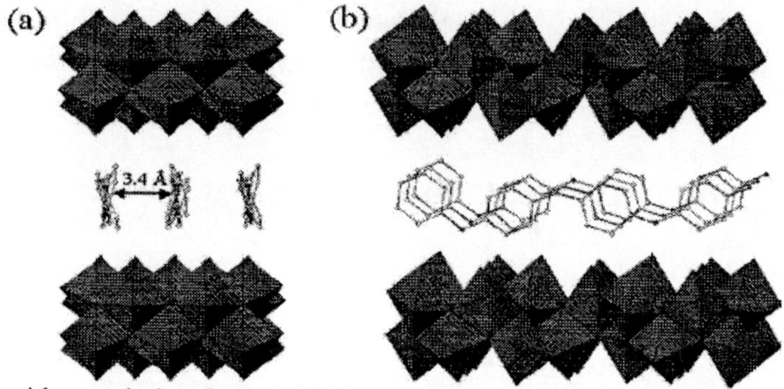

Reproduced with permission from ACS "Three-Dimensional Structure of Nanocomposites from Atomic Pair Distribution Function Analysis: Study of Polyaniline and (Polyaniline)0.5V2O5.1.0H2O" V.Petkov, V.Parvanov, P.Trikalitis, C.Malliakas, T.Vogt, M.G.Kanatzidis ,J.Am.Chem .Soc., 2005, 127:8805-8810.

Figure 51. Atomic ordering in $(PANI)_{0.5}V_2O_5{\bullet}1.0H_2O$ nanocomposite along (a) a and (b) b axis of the monoclinic lattice used to describe the inorganic V_2O_5 matrix.

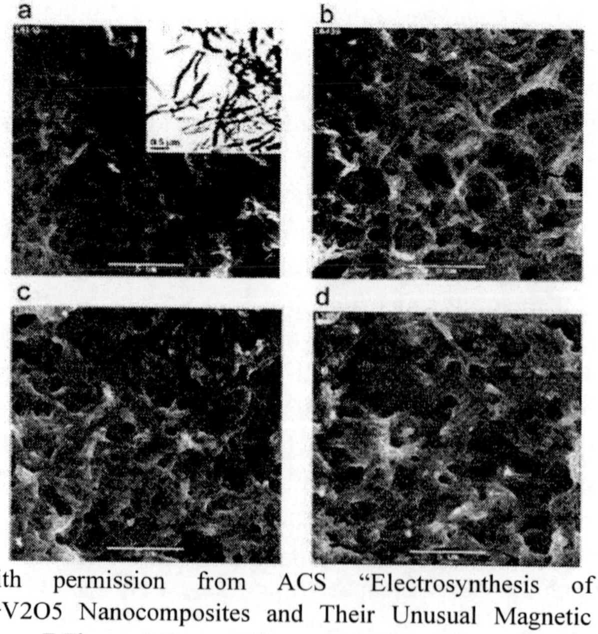

Reproduced with permission from ACS "Electrosynthesis of Macroporous Polyaniline−V2O5 Nanocomposites and Their Unusual Magnetic Properties", I. Karatchevtseva, Z.Zhang, J.Hanna, V.Luca, and Chem.Mater. 2006, 18, 4908-4912.

Figure 52. SEM and TEM (inset on a) images of as-prepared PANI film (a) and PANI-V_2O_5 nanocomposites deposited for 30 min as a function of current densities of (b) 1, (c) 3 and (d) 5 mA/cm^2. The scale bar is 5 nm in length.

The thickness of the PANI-V_2O_5 nanosheets was found to decrease with an increase with the aniline concentration, Figure 51 Starting from mesoporous V_2O_5 containing a small amount of Mg, PANI-V_2O_5 nanocomposites with lamellar and wormlike structures and fibrous morphologies were synthesized by an in-situ oxidative intercalation polymerization of aniline in air with V_2O_5 being the oxidation agent [232]. The resulting composite kept an ordered mesostructure. The PANI chains were intercalated into the mesostructured V_2O_5/Mg. The conductivity of the PANI-V_2O_5 composites reached values of the order of 10^{-2} S/cm., being 5 times higher than that of mesostructured V_2O_5/Mg. *Karatchevtseva et al.* [233] fabricated interpenetrating PANI-V_2O_5 nanocomposites starting with a spongy interconnected PANI network electrodeposited on a titanium metal ,Figure 52 .The network, with pores of the order of a few micrometers, was used as a template for V_2O_5, electrochemically deposited as well. The control of electrochemical parameters gives some control over the thickness of the V_2O_5 component, and, hence, the porosity. PANI appears to remain in its most conductive state if the current density used for depositing V_2O_5 is lower than 2 mA/cm^2. *Malta et al.* [234-235] synthesized PANI-V_2O_5 nanofibres and nanotubes. The synthesis route development by *Nesper et al.* at the *ETH Zürich* was the base for the production of nanofibres [236-237] and nanotubes [238]. The PANI-V_2O_5 nanofibres were obtained as the main reaction products of (PANI$_{0.3}$)–V_2O_5 powder and hexadecylamine after hydrothermal treatment, and had typical lengths of 1 to 10 nm and a width varying from 15 to 400 nm. Electrochemical experiments showed a specific capacity of about 150 Ah/kg.The PANI-V_2O_5 nanotubes were produced by a cationic exchange between hexadecylamine and polyaniline after synthesis of V_2O_5 nanotubes by a sol-gel method followed by hydrothermal treatment. Polyaniline was found to be intercalated in the interlamellar space of the PANI-V_2O_5 nanotubes forming a hybrid material with well defined morphology. These nanostructures improved the diffusion kinetics of lithium intercalation. *Cuentas-Gallegos and Gomez-Romero* [239] studied the integration of an electro-active inorganic molecule into a conducting polymer (PANI or PPy) that in turn was intercalated into an extended inorganic oxide (V_2O_5) to fabricate triple hybrid nanocomposites (or ternary nanocomposites). Alternatively, *Posudievsky et al.* [240] obtained triple hybrid nanocomposites based on V_2O_5 in the interlayer galleries of which the macromolecules of an ion-conducting polymer (PEO) and an electron-conducting polymer (PANI).The synthesis of PANI-HCF-V_2O_5 triple hybrid material was carried out via an oxidative polymerization of aniline, in the presence of $H_3Fe(CN)_6$. The two inorganic reagents oxidized aniline to get incorporated into the hybrid compensating at the same time the positive charge of the conducting polymer. The preparation of the PANI-PEO-V_2O_5 nanocomposite was realized by the insertion of PANI into the structure of a preliminarily prepared PEO-V_2O_5 nanocomposite, either by in-situ intercalative polymerization of aniline or by direct intercalation of PANI macromolecules. Repeated charge-discharge cycling resulted for the PANI-HCF-V_2O_5 nanocomposite in poor cyclability, and for the PANI-PEO-V_2O_5 nanocomposite in improved charge capacity and stability. Another novel ternary nanocomposite was generated from the redox polymerization of aniline inside a lamellar betonite-vanadium [V] oxide [BV] matrix [241-242]. The BV matrix is prepared by combining bentonite and polyvanadic acid gel. Bentonite is a clay material usually referred to as sodium montmorillonite.The presence of intercalated polymer seems to stabilize the BV network, minimizing the structural reorganisation usually required for the insertion of lithium into the matrix. Interestingly, in small amounts, PANI dramatically increased the conductivity and the charge capacity of the BV matrix, whereas increasing amounts of PANI led to an opposing effect. A nanocomposite based on PANI, 12-phosphormolibdic acid [PMA], and V_2O_5 was prepared by *Posudievsky et al.* [243] via direct intercalation of the macromolecules of the secondary doped conducting polymer inside the interlayer galleries of the oxide.The features of the preparation method - especially, the use of *m*-cresol as a solvent for PANI, as a secondary dopant was reflected in the electro-physical and paramagnetic characteristics of the nanocomposite. The PANI•PMA-V_2O_5 composite possesses a sufficiently high conductivity (5.0 x 10^{-2} S/cm) and is an effective electro catalyst of oxygen reduction in weak acid solutions, important in, for instance, zinc-air batteries. *Rajesh et al.* [244-245] used the Pt nanoparticles loaded PANI-V_2O_5 nanocomposite, prepared via the intercalation in the V_2O_5 matrix, as the catalyst support for Pt nanoparticles.The nanocomposite exhibited excellent electrochemical and thermal stability in comparison to the physical mixture of PANI and the transition metal oxide. The nanocomposite showed high electro catalytic activity and stability for methanol oxidation compared to the commercial, Pt loaded Vulcan CX 72 carbon support. The higher activity and stability are attributed to the better CO tolerance of the composite material.

6.3.2. PANI-TiO_2 Nanocomposites

In-situ polymerization of aniline is often used to synthesize PANI-TiO_2 nanocomposites, with ammonium peroxodisulfate [APS] as oxidant [246-251] and iron sulphate [247] was used additionally. TiO_2 clusters are broken down under ultrasonic action prior to the aniline polymerization, while during polymerization continuous mechanical stirring prevents re-agglomeration; in [252] in-situ polymerization is performed by adding $(NH_4)_2S_2O_8$ to a mixture of aniline hydrochloride solution and TiO_2 sol-gel. In order to improve the processability, the emeraldine salt of the nanocomposite, synthesized according the to indicated route and was converted in its base form by protonation with several types of sulfonic acids [246]. The conductivity of doped PANI-TiO_2 prepared from *m*-cresol suspension was higher that from chloroform suspension. Homogeneous suspensions in chloroform of SA-doped PANI solution and anionic surfactant-coated TiO_2 resulted in improved processability and a conductivity of 1 to 10 $S.cm^{-1}$ for cast films.Spectroanalysis of PANI-TiO_2 nanocomposites of rutile [247] and anatase [248] TiO_2 showed that PANI macromolecules and TiO_2 nano-particles are not simply mixed but are strongly interacting. Synthesized PANI was found deposited on the surface of TiO_2 particles forming a core-shell structure [248]. The strong interaction is attributed to the formation of H-bonding, Figure 53.

Reproduced with permission from Elsevier, "Titanium dioxide doped polyaniline", J-C Xu, W-M Liu, H-L Li, Mater Sci Eng C, 2005, 25,444-448.

Figure 53. SEM images of in-situ polymerized PANI-TiO_2 composites with different loadings of TiO_2, (a) 5, (b) 25, (c) 50 and (d) 80 wt %.

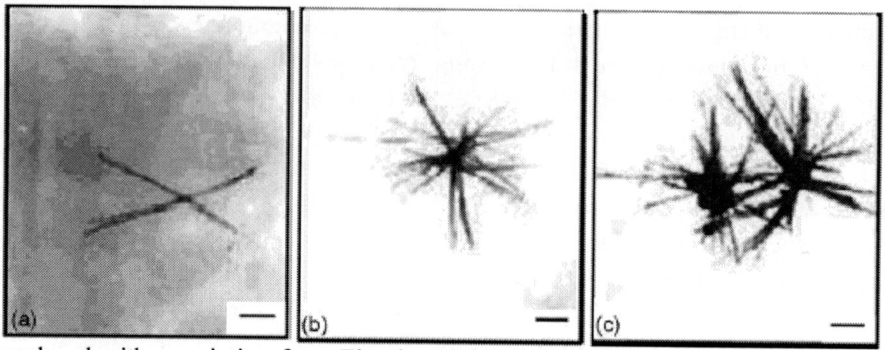

Reproduced with permission from Elsevier "Self" X. Sui, Y.Chu, S.Xing, M.Yu, C.Liu, Colloids Surf A , 2004, 251,103-109.

Figure 54. TEM images of PANI-TiO_2 synthesized in a OP/hexamethylene/ water system.

PANI-TiO_2 composites with different organic/inorganic ratios could be degraded by photocatalytic oxidation, showing that such composite have a potential to be used as a photodegradable product [250].PVA-TiO_2 membranes were prepared by incorporating nanosized TiO_2 particles and PANI-TiO_2 composite into PVA and crosslinking with glutaraldehyde. The PANI-TiO_2-PVA membrane was superior in pervaporation separation of water-isopropanol mixtures [251]. PANI-TiO_2 composite film showed rapid response and good reproducibility to trimethylamine, and good selectivity to analogous gases [252].Fine grade powder (particle size ca. 100 nm) of anatase TiO_2 was used in the polymerization mixture, and a PANI-TiO_2 composite was obtained [253]. A much higher yield of PANI was found to be formed in the presence of TiO_2 particles, than without TiO_2 added. The composites obtained exhibited high piezoresistivity with piezosensitivity having a maximum value at a definite composition of a composite. Alternatively, spherical PANI-TiO_2 nanoparticles were successfully synthesized by the oxidation of precursor with APS in reverse micelles with OP, AOT and CTAB serving as surfactant [254]. Reverse micelles provided control of the unique structure and morphologies of nanocomposites, Figure 54. After 60 days, the spherical nanoparticles were self-organized to sea urchin-like shape.

PANI-TiO_2, PANI-AgCl and PANI-$BaSO_4$ nanocomposites were also synthesized in reverse micelles but now with CTAB as surfactant [255].PANI-TiO_2 nanocomposites are prepared by in-situ polymerization with APS in a TiO_2 colloid under sonication [256]. The dc-conductivity does not change significantly with PANI concentration. The ac-conductivity of these nanocomposites reveals a correlated barrier hopping type conduction process. Very high dielectic constants were found at room temperature.PANI-TiO_2 bilayer microtubes with a diameter of 200 nm were prepared on anodic aluminium oxide [AAO] template in a two-step method [257]. Firstly, PANI microtubes were synthesized in the channels of the template by in-situ polymerization within a solution of APS and HCl. Secondly, the TiO_2 microtubes were synthesized with sol-gel method on the inner surfaces of the PANI microtubes. The photo-catalysis efficiency of PANI-TiO_2 microtubes is higher than that of TiO_2 microtubes, since PANI can act as sensitizer of TiO_2.

6.3.3. PANI-ferrite Nanocomposites

Nanocomposites/nanoparticles of magnetic iron oxides, such as Fe_2O_3 and Fe_3O_4, and the conducting polymer PANI, are studied because of the interesting combination of properties and potential applications in such diverse areas as information storage, colour imaging, ferro-fluids, medical diagnosis and controlled drug delivery. Chemical and mechanical routes are pursued to prepare the nanocomposites/nanoparticles. The most general, chemical method to prepare iron oxides / PANI nanocomposites is by in-situ polymerization of aniline monomers in an aqueous solution in the presence of dispersed iron oxide particles. However, *Lu et al.* [258] used a novel method: aniline dimer-COOH assisted polymerization,Figure 55.

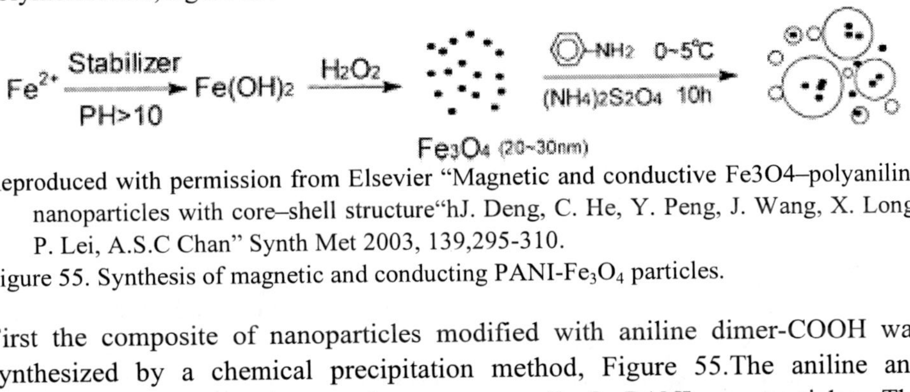

Reproduced with permission from Elsevier "Magnetic and conductive Fe3O4–polyaniline nanoparticles with core–shell structure"hJ. Deng, C. He, Y. Peng, J. Wang, X. Long, P. Lei, A.S.C Chan" Synth Met 2003, 139,295-310.

Figure 55. Synthesis of magnetic and conducting PANI-Fe_3O_4 particles.

First the composite of nanoparticles modified with aniline dimer-COOH was synthesized by a chemical precipitation method, Figure 55.The aniline and oxidant were added subsequently to prepare Fe_3O_4-PANI nanoparticles. The Fe_3O_4 nanoparticles were well dispersed in the PANI matrix, the Fe_3O_4-PANI nanoparticles were found to be paramagnetic. At room temperature, they possessed high saturation magnetization (M_s = 21 emu/g) and a low coercive force (~ 0). *Ashesteguy and Jacobo* [259] controlled the pH during the entire in-situ polymerization process. They found the electrical and magnetic properties to depend strongly on the pH value at room temperature. Ferromagnetic properties were observed in samples prepared with pH > 4 (M_s = 80.4 emu/g, coercive force ~ 0 at pH = 11.5). The magnetic particles appeared to improve the thermal stability of the nanocomposite due to interactions between the particles and the PANI chains. Deng et al. [260] reported the preparation of PANI-Fe_3O_4 nanoparticles with core-shell structure via in-situ polymerization of aniline monomer in an aqueous solution, which contains Fe_3O_4 nanoparticles and surfactant dodecylbenzene sulfonic acid sodium salt [NaDS]. The nanoparticles had an average size of about 80 nm, with cores of Fe_3O_4 nanoparticles ranging from 20-30 nm and a conductive shell composed of aniline,Figure 55.The value of saturated magnetization and coercive force depend on the Fe_3O_4 content (M_s = 4.2 - 48.4 emu/g; H_c = 8-55.3 Oe).

Long et al. [261] prepared Fe_3O_4-(NSA-PANI) nanotubes (~150 nm in diameter) and nanorods (~140 nm in diameter), containing Fe_3O_4 nanoparticles of about 12 nm, by the template-free method in the presence of NSA as dopant. The conductivity of the nanostructures was 10^{-1} to 10^{-2} S/cm. The nanocomposites (6 and 20 wt % of Fe_3O_4) show a large negative magneto resistance compared with pure PANI nanotubes and a considerably lower saturation magnetization (M_s = 3.45 emu/g at 300 K and 4.21 emu/g at 4 K) compared with the values measured for pure Fe_3O_4 nanoparticles (M_s = 65 emu/g).Besides chemical preparation routes mechanical/physical routes are explored as well. *Xue et al.* [262] mixed DBSA-PANI powder and (HCl-PANI)-Fe_3O_4 powder mechanically in an agate mortar. The conductivity of the nanocomposites decreases linearly from 0.19±0.06 to 0.05±0.06 S/cm when the (HCl-PANI)-Fe_3O_4 content was increased from 0 to 100 wt %. The conductivity varied with temperature, revealing that the charge transport mechanism can be considered to be one-dimensional variable range hopping. All the PANI-Fe_3O_4 nanocomposites showed magnetization curves without a characteristic hysteresis.*Bao and Jiang* [263] used high-energy ball milling to prepare PANI-Fe_3O_4 mixtures, comprising 50 wt % Fe_3O_4 and 50 wt % PANI. Increasing milling time reduced the size of the Fe_3O_4 particles rapidly to about 21 nm. Broken PANI chains reacted with the Fe atoms, forming some paramagnetic products and a small number of super magnetic γ-Fe_3O_4 particles. *Lu et al.* [264] prepared PANI nanotubes containing Fe_3O_4 nanoparticles under ultrasonic irradiation of aqueous solutions of aniline, ammonium peroxydisulfate, phosphoric acid and the quantitative amount of aniline dimer-COOH capped Fe_3O_4 nanoparticles,Figure 56.TEM images revealed nanotubes and proved that Fe_3O_4 nanoparticles were embedded in the composites. The nanotubes had a super magnetic behaviour.

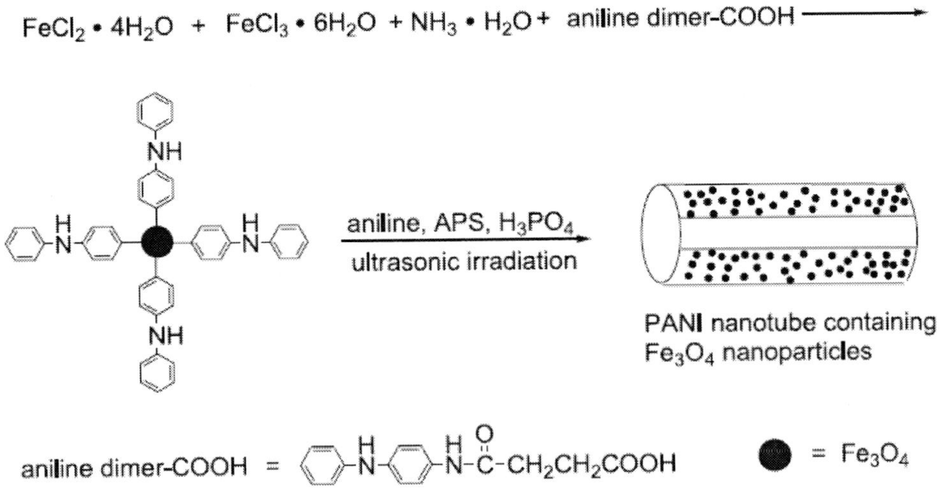

Reproduced with permission from Elsevier "Ultrasonic synthesis, X. Lu, H. Mao, D. Chao, W .Zhang, Y. Wei , J Solid State Chem 2006,179,2609-2615.

Figure 56. Fe_3O_4 - PANI nanotubes synthesis assisted by ultrasonic irradiation.

Tang et al. [265] studied the preparation of (γ-Fe_2O_3)-PANI nanocomposite films. Their approach of employing anionic surfactants enabled coating/stabilization of the (γ-Fe_2O_3)-PANI nanoparticles with ω-methoxypoly(ethylene glycol) phosphate [PEOPA], 4-dodecyl-benzenesulfonic acid [DBSA], and 10-camphorsulfonic acid [CSA], dissolution of the PANI chains, a doping of the emeraldine base without acids as HCl, amalgation of γ - Fe_2O_3 and PANI in common solvents, (5) a fabrication of free-standing nanocomposite films by static casting, and (6) a manipulation of the γ-Fe_2O_3 content and film thickness by changing the (γ-Fe_2O_3) / PANI ratio and concentration. The (γ-Fe_2O_3)- (DBSA-CSA-PANI) nanocomposite films show electrical conductivity of 82-237 S/cm. The nanocomposite films are superparamagnetic, showing no hysteresis at 300 K and exhibiting a blocking temperature in the temperature region of 63-83 K. *Yang et al.* [266] prepared (γ-Fe_2O_3)-PANI nanocomposite by the micro-emulsion method. To improve its electrical properties, the nanocomposite was doped by emulsion-extraction again in the presence of γ-Fe_2O_3 nanoparticles. The products are not only transparent, but also have good quality in conducting and magnetic properties. Reverse micelle microemulsion was employed to prepare magnetic and conductive PANI-NiZn ferrite nanocomposites [267]. A stable NiZn ferrite magnetic fluid was obtained from boiling an aqueous $Ni_{0.5}Zn_{0.5}Fe_2O_4$ nanoparticles colloid suspension in an alkaline solution. Aniline was polymerized in the mixture of the colloid suspension and the microemulsion made from Triton X-100/*n*-hexanol/*n*-hexane solution and an aniline/HCl mixture when ammonium persulfate was added. The obtained PANI-NiZn ferrite nanocomposites displayed a core-shell structure with particles ranging from 30 to 50 nm; the core of NiZn ferrite nanoparticles had an average particle size of 12 nm. Coating of the particles with PANI significantly affected their conductive and magnetic properties: conductivity and blocking temperature decreased, and coercivity was almost unchanged, while the saturation magnetization drastically decreased. Polycrystalline MnZn ferrite was coated with PANI and embedded in a polyurethane matrix [268]. Grinded down particles of sintered MnZn ferrite were immersed in an aqueous solution of aniline hydrochloride followed by an oxidation polymerization and resulting in coating of the particles with a film of 100-200 nm thickness. The PANI coating of the particles was deprotonated in ammonium hydroxide, and protonated in aqueous solutions of picric acid. By adjusting the acid concentration, the degree of protonation of the PANI coating and, consequently, also its conductivity was controlled over several orders of magnitude. Large shifts in the resonance frequency in the frequency range of 1 MHz to 3 GHz were observed as a function of PANI conductivity.

6.3.4. PANI- Al_2O_3

The chemical oxidative polymerization of a mixture of *N*-vinylcarbazole [NVC] and aniline, via $FeCl_3$ in aqueous medium, produced an insoluble polymer that contained PANI along with some residual poly-*N*-vinylcarbazole [PNVC] not extractable with benzene [269]. This insoluble PANI-PNVC composite could be obtained as a stable dispersion in water when the above polymerization was conducted in the presence of nanosized Al_2O_3. Alternatively, (PANI-Al_2O_3)-PNVC composite could also be produced by precipitating a THF solution of preformed PNVC onto a PANI-Al_2O_3 suspension. Likewise, PANI-(PNVC-Al_2O_3) could be obtained by polymerization of aniline onto an aqueous suspension of PNVC-Al_2O_3. The inclusion of either of these monomer units in the resulting polymer was confirmed by FTIR analysis. SEM analyses revealed the formation of a lumpy aggregate, and TEM images revealed particles sizes of the (PANI-Al_2O_3)-PNVC composite to be in the 52-200 nm range.

6.3.5. PANI-SnO_2 Nanoparticles

A PANI-SnO_2 hybrid material was prepared by a thermal method [270]. Aniline was added to a water-HCl mixture (1 mol/L), and after ultrasonication,

SnO_2 was added. The mixture was stirred and ultrasonicated for 30 min, before being transferred to a Teflon autoclave, where the addition of ammonium peroxodisulfate aniline at a molar rate to aniline of 1:1 followed. The autoclave was sealed and kept at 140 °C for 4 h.XRD data showed that the crystal structure of SnO_2 was not changed by PANI, but SnO_2 affected the crystallization to some extent. PANI and SnO_2 were found to be dispersed uniformly, but not all the SnO_2 particles were totally covered by PANI, Figure 57.The PANI-SnO_2 hybrid material was found to be sensitive for both ethanol and acetone, when operated at 60 or 90°C and showed good reversibility. When operated at 90°C, the hybrid material exhibited response and recovery times within 1 min.The sensing mechanism could be related to the existence of p-n hetero junctions in the PANI-SnO_2 hybrid material.

Reproduced with permission from Elsevier "Characterization and gas sensitivity study of polyaniline/SnO2 hybrid material prepared by hydrothermal route" L .Geng, Y. Zhao, X .Huang, S .Wang, S .Zhang, S. Wu ,Sens Actuators B 2007, 120,568-573.

Figure 57. HRTEM micrograph of PANI-SiO_2 hybrid material.

6.3.6. PANI- PANI-ZnO and PANI-CeO_2 Nanocomposite

The PANI-(nano-ZnO) composite particles could formed stable hollow spheres by assembling at the surfaces of the droplets in the Pickering emulsion [271]. Their diameter decreased with the increasing amount of ZnO nanoparticles in the reaction chamber. When the volume ratio of toluene to water (R) was in the range of 0.03 to 0.07 [272], stable Pickering emulsions was formed and the oxidative polymerization of aniline resulted in PANI-(nano-ZnO) fibres; when R

> 0.07, the volume fraction of the dispersed phase was too large to form stable Pickering emulsion, and at the continuously changing toluene-water interface dendrite PANI-(nano-ZnO) composites were formed. When R < 0.03, there was no toluene-water interface anymore and the PANI-(nano-ZnO) composites became polyhedral particles through accretion. *He et al.* [271-272] prepared PANI-CeO_2 microspheres, and PANI-(nano-ZnO) composites and composite fibres in pickering emulsions stabilized solely by CeO_2 and ZnO nanoparticles. Nanoparticles, aniline, toluene, ammonium peroxodisulfate and distilled water were introduced in the reaction vessel under magnetic agitation. In case of PANI-(nano-ZnO) composite fibres, 10% acetic acid was added to the water at the ratio of 1:50. The emulsion was ultrasonicated: in the preparation of PANI-CeO_2 microspheres for 2 h at 25°C, and in the preparation of PANI-(nano-ZnO) for 1 h at 20°C. The PANI-CeO_2 composite microspheres [273] comprised flaky composite particles. In the composites PANI was amorphous, while the CeO_2 nanoparticles maintained their cubic crystal structure. The nanoparticles were partly enwrapped by PANI.

6.3.7. PANI-MnO_2 Nanocomposites

PANI-MnO_2 composites were chemically prepared by oxidative polymerization of aniline in acidic media containing MnO_2 as oxidant, Figure 58 [274]. The composites were produced by injecting liquid aniline into a polymerization vessel containing MnO_2 suspended in1.14 mol/L of aqueous acids solutions (H_2SO_4, HNO_3, HCl, and H_3PO_4) at room temperature. The polymerization of aniline was found to be initiated immediately after the above addition and the mixture was stirred for 1 h. XRD measurements of the composites showed that the crystal structure of incorporated MnO_2 was distorted and converted into an amorphous state. The degree of crystallinity of the prepared composites is dependant on the type of the compensated acid anions and has the order $HSO_4^- > H_2PO_4^- > Cl^- > NO_3^-$.

From the values of point zero charge of MnO_2 (pzc at pH ≈7.5) and FTIR spectra it was concluded that the absorbed acids anions on the oxide surface works as the charge compensator for positively charged PANI in the formation of PANI-MnO_2.Magnetic susceptibility measurements revealed that the composites had diamagnetic properties.

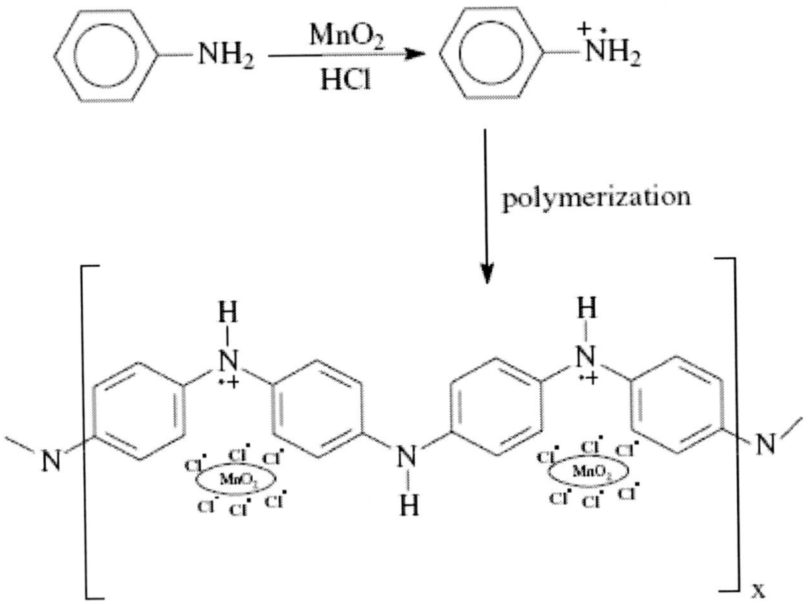

Reproduced with permission from Elsevier "Preparation and characterization of polyaniline/manganese dioxide composites via oxidative polymerization: Effect of acids " A.H. Gemeay, I.A. Mansour, R.G. El-Sharkawy, A.B. Zaki, Eur Polym J 2005, 41,2575-2583.

Figure 58. PANI-MnO_2 composite, made by polymerization of aniline in aqueous acidic (HCl) medium with MnO_2 as oxidant.

6.3.8. PANI-phosphate Nanocomposites

De Castro et al. [275] prepared a novel flexible, free-standing and transparent organic/inorganic hybrid materials formed by a procedure in which the PANI (about 2.42 wt %) in presence of a polyphosphate inorganic host structure (arising from a sol–gel transition) Figure 59.The conducting polymer, polyaniline in its green conducting emeraldine salt form, was entrapped within the gel network. The polymer was formed in small domains dispersed in the inorganic matrix.

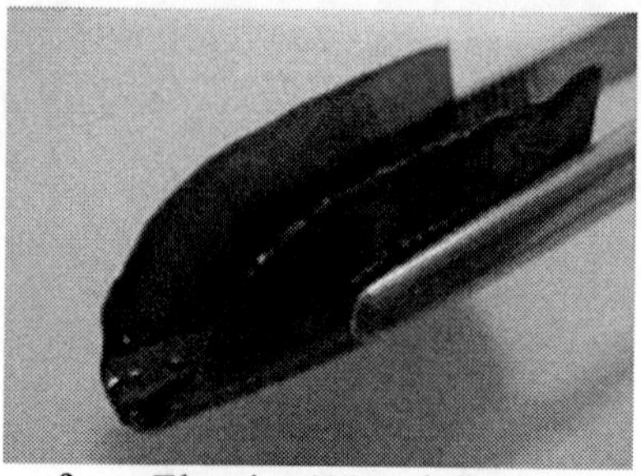

Reproduced with permission form Elsevier "Novel flexible, freestanding and transparent organic/inorganic hybrid materials formed between polyaniline and polyphosphate gel" E.G. De Castro, A.J.G Zarbin, H.P. De Oliveira, A .Galembeck ,Synth Met, 2004,146,57-62.

Figure 59. Green tinted polyaniline / aluminium polyphosphate [PANI-APP] gel composite.

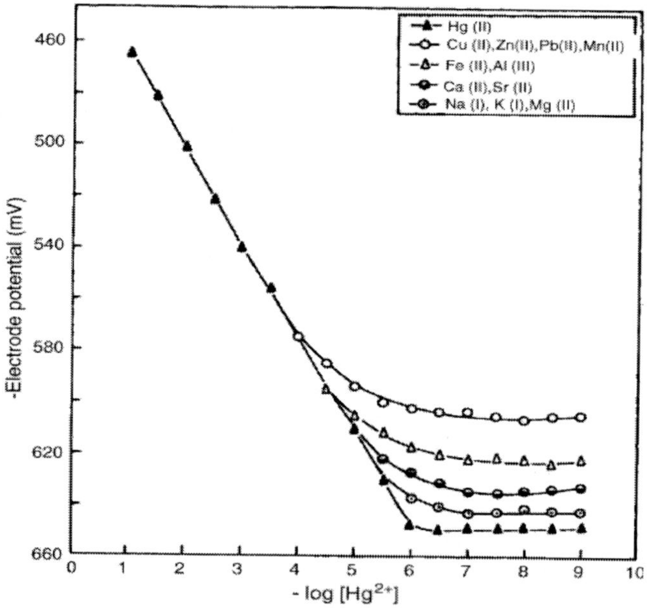

Reproduced with permission from Elsevier "Applications of Hg(II) sensitive polyaniline Sn(IV) phosphate composite cation-exchange material in determination of Hg2+ from aqueous solutions and in making ion-selective membrane electrode" A.A.Khan , Inamuddin , Sens Actuators B,2006, 120,10-18.

Figure 60. Selectivity coefficients of various interfering ions for the PANI Sn (IV) phosphate membrane electrode.

Since the composites were transparent, the PANI domains are small enough to scatter visible light; these hybrids formed true nanocomposites. PANI Sn(IV) phosphate, an 'organic-inorganic' composite material, was prepared via sol-gel mixing of PANI into the matrices of the inorganic precipitate of Sn (IV) phosphate [276]. The ion-exchange capacity and distribution coefficients with respect to various metal ions were determined on the column of PANI Sn(IV) phosphate. The distribution studies showed the selectivity of this material for Hg (II) ions. A membrane electrode, made of PANI Sn(IV) phosphate, showed a high sensitivity, selectivity and fast potential response to Hg (II) ions. The membrane electrode could determine Hg (II) ions in a pH range of 4-9 with a slope of 30 mV per decade change in a concentration range of $1x10^{-1}$ to $1x10^{-6}$ M. The electrode showed good reproducibility, detection limit and a lifetime of 3 months. Selectivity coefficients determined by the mixed solution method, revealed that the electrode was selective for Hg (II) ions in the presence of interfering cations, Figure 60.

Zirconium phosphate sulfophenyl phosphate was employed as a template for aniline because of the presence of strongly hydrophilic sulfo groups, and the resulting molecular complex was polymerized to PANI [277]. Metal phosphonates $Ti(HOPO_3) \bullet H_2O$, γ-$Zr(HOPO_3)_2 \bullet H_2O$, and $HUO_2PO_4 \bullet 4\ H_2O$ were used for intercalation of aniline and N-phenyl-p-phenylenediamine, and, after thermal treatment at 130 °C in air, intercalated aniline and its derivative were slowly polymerized by ambient oxygen [278]. Following polymerization, the interlayer spacing increased slightly from 10.46 to 11.8 Å for anilinium/uranyl phosphate, but decreased greatly, from 28.3 to 20.5 Å, for zirconium phosphate, and from 18.87 to 13.25 Å for titanium phosphate, indicating a monolayer of PANI inside the uranyl and titanium phosphates, while a bilayer in zirconium phosphate. High intensity ultrasound (20 kHz, ~100 W/cm^2, for several hours) was applied to prepare nanocomposites of PANI and γ-zirconium phosphate $\tilde{\gamma}$-ZrP [279]. The intercalation rate was greatly enhanced, and the interlayer distance of aniline-intercalated ZrP was found to be 16.0 Å. During sonication the intercalated aniline polymerized at low pH after addition of ammonium peroxodisulfate. Nanocomposites of $\tilde{\gamma}$-ZrP absorbed and intercalated by PANI that incorporated amorphous Cu_2O and nanocrystalline CuO, were also synthesized by a sonochemical method [280]. Cu^{2+} was pre-intercalated into $\tilde{\gamma}$-ZrP to form layered γ-CuZrP host that provided a template for the polymerization of aniline. X-ray diffraction patterns for the nanocomposite indicated intercalation of doped PANI in $\tilde{\gamma}$-ZrP. Nanocomposites of PANI and γ -zirconium phosphate [$\tilde{\gamma}$-ZrP] were prepared, based on adsorption [281]. Aniline was mixed with water, $\tilde{\gamma}$ -ZrP was added and the mixture was stirred at room temperature for 24 h using a magnetic stirrer. X-ray diffraction patterns showed a lattice expansion from 7.6 to 14.1 Å. It is interesting to note that aniline forms mono- and bilayers in $\tilde{\gamma}$-ZrP. The encapsulated aniline was polymerized by adding ammonium peroxodisulfate, reducing the interlayer distance to 7.8 Å. This indicated polymerization of aniline outside the interlayer of crystalline $\tilde{\gamma}$-ZrP. The polymers of the intercalated monomer composite were more ordered than those formed directly after mixing of monomer and γ-ZrP and revealed quite different dc and ac electrical properties than the latter. The charge transport phenomena are interpreted by electron and polaron tunnelling for the intercalated monomer systems.

6.3.9. PANI-CdS Nanocomposites

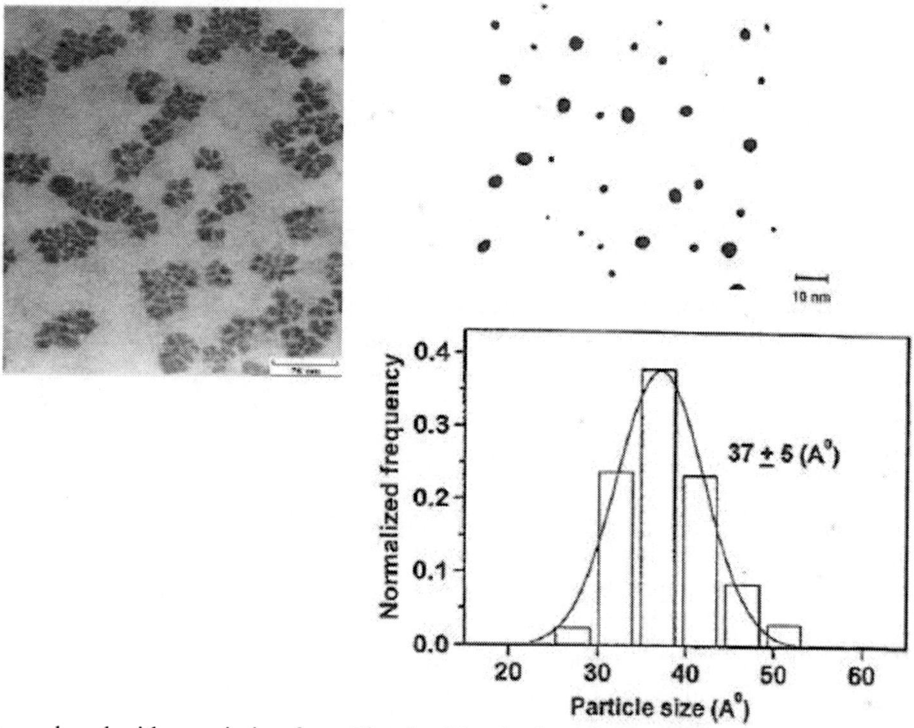

Reproduced with permission from Elsevier "Synthesis and characterization of conducting polyaniline-coated cadmium sulphide nanocomposites in reverse microemulsion"P.S. Khiew, N.M.Huang, S.Radiman , A. Md. Soot Ahmad, Mater Lett, 2004,58,516-521.

Figure 61. TEM image of particles in PANI-CdS composite film (b) particle size distribution histogram.

PANI-CdS composite films were electrochemically deposited on fluorine-doped tin oxide coated glass substrates from electrolyte containing mixtures, with different molar ratios, of CdS particle and aniline in an acidic medium [282]. The average particle size in the electrolyte was about 30 Å and no considerable increase was observed after film formation. Films with good optoelectronic properties could be made by controlling monomer concentration, applied potential (+0.7 V) and pH (3.5 – 4) of the electrolyte.

Khanna et al. [283-284] synthesized nanocomposites of PANI and CdS via in-situ generation of CdS from highly reactive organometallic cadmium precursor, i.e. dimethyl cadmium in *N,N'*-dimethyl formamide [DMF] [284].The bubbling of H_2S through the mixture instantly resulted in a colour change of deep blue to green. The composites showed a mixture of agglomerated and non-agglomerated fluffy particles. PANI-CdS nanocomposites were alternatively prepared in a water/oil non-ionic microemulsion system, Figure 61 [285]. In an alternative method, potassium peroxydisulfate was used for the polymerization of aniline, cadmium nitrate and sodium sulphide-hydrate to synthesize nanocrystalline CdS and cyclohexane as theoil phase together with a surfactant consisting of poly $(oxyethylene)_5$ nonyl phenol ether and poly $(oxyethylene)_{10}$ nonyl phenol ether in the weight ratio of 2:1. The PANI-CdS nanocomposites were polydispersed and had an average size of about 17.8 nm, Figure 61. The intralayer molecular packing in PANI-CdS showed hexagonal layer plane packing with the original rectangular (herringbone) packing. PANI-CdS composite films were prepared, using metal trifluoromethanesulfonates as NMP-soluble precursors for the in-situ preparation of metal sulphides [286-287]. UV/VIS absorption spectra suggested that the PANI-CdS nanocomposites were stable in air, while PANI was found to degrade in the presence of Cu_2S. Both short-circuit currents and open circuit voltages of photovoltaic devices made with PANI-CdS were found to increase with the concentration of CdS in the polymer matrix.

6.3.10. PANI-TiN Nanocomposites

PANI-TiN nanocomposites have been prepared by in-situ chemical polymerization of aniline in the presence of TiN nanoparticles dispersed by sonication as described by *Qiu and Gao* [288].The nanocrystalline TiN with an average diameter of ~20 nm was incorporated and dispersed homogeneously within the polymer matrix, leading to enhanced conductivity and electrochemical activity. The morphology and structure of the PANI-TiN was controlled by the content of TiN nanoparticles in the composites. When the TiN content was low,

TiN nanoparticles were embedded in a featureless amorphous polymer matrix. When the load of TiN nanoparticles was as high as 30 wt %, TEM, Figure 62, revealed that the PANI particles began to grow into an individual rod like structure with a diameter of 100-300 nm and a length up to a micrometer. At a higher TiN content of 50 wt %, most of the TiN nanoparticles were incorporated into composite rods. These nanocomposite rods agglomerated side-by-side with each other, as shown in Figure 62. The TiN nanoparticles were embedded in the composite rods and, at the same time, aligned spontaneously along the edges of the rods.

Qiu Y, Gao L (2005) J Phys Chem B 109:19732.
Reproduced with permission form ACS "Novel Polyaniline/Titanium Nitride Nanocomposite: Controllable Structures and Electrical/Electrochemical Properties" Y Qiu, L Gao () J Phys Chem B 2005,109,19732-19744.

Figure 62. TEM photograph agglomerated PANI-TiN (50 wt %) composite rods.

The electrical and electrochemical properties of the nanocomposites were also affected by the structure. In these composites, the PANI chains grow into integrated homogeneous composite particles in the presence of TiN nanoparticles, which cause the decrease of grain boundary area as well as the resistivity.

6.3.11. PANI-Zeolite

First attempts by *Bein and Enzel* [289-290] to incorporate PANI into channels of zeolites (three-dimensional channel system) and mordenite (one-0dimensional channel system) as open framework oxide structures with pore dimensions between 0.3 and 0.8 nm did not lead to any measurable functionality (conductivity) of the resulting nanocomposites. The PANI was stabilized inside the zeolite channel system, however, the polymerization rate was found to be orders of magnitude slower in the zeolite compared to the bulk reaction. Polymerization took place in the presence of zeolitic protons and no evidence of bulk polymer covering the outer surface of the PANI / zeolite system was found [290]. The degree of conversion was approximately 15 % as determined by dissolution of the zeolite systems in HF.A subsequent study by *Frisch et al.* [291] concerning the magnetic properties of PANI /zeolite nanocomposite materials resulted in hybrids which showed an appreciable antiferroelectric component at weight ratios PANI /zeolite larger than 20. This novel antiferromagnetism was speculated to arise from π-dimer pairing of PANI chain layers stacking on the zeolite particle surface. A retardation of the aniline conversion by the presence of the channel network was observed, and significantly lower values of Mw for the PANI compared with bulk polymerization at excessive load of aniline with somewhat slightly higher polydispersity were obtained for samples with 1:1 wt % ratios. This indicated that the presence of the zeolite surface somehow impedes diffusion of monomer thereby impeding the synthesis. The conductivity levels were observed to start at less than 10^{-8} S/cm for PANI/zeolite ratios of 1:1 up to 1.2 S/cm for a ratio 19:1. Electrical conductivity response of PANI / zeolite nanocomposites towards CO was investigated by *Chuapradit et al.* [106] in terms of zeolite content of the composite and zeolite pore size. The nanocomposites were prepared by dry mixing of PANI powder with the zeolite powder followed by compression moulding. The addition of zeolite 4^0A and 5^0A reduces the electrical conductivity response but improves the sensitivity towards CO with increasing zeolite concentration (up to 40 wt %). This concentration is below the percolation threshold value (estimated above 50 wt %). The composites using zeolite 3^0A display a comparable sensitivity value relative to the bulk PANI. Zeolite 5A is the most effective since it displays the largest pore sizes. In-situ oxidative adsorption polymerization of aniline vapour in Cu- and Fe-loaded

MCM-41 nanoporous materials was first reported by *Wu and Bein* [292]. A typical polymer loading of 0.16 g PANI / g MCM has been achieved with a typical M_w of the interchannel PANI of 40000. The diffusional constraints experienced by the reactants limited the chain growth within the MCM channels. The stability of the emeraldine salt against oxygen was increased due to the presence of the host. The disappearance of the Si-OH peak in the Raman spectra hinted to a complete coverage of the intramolecular channels with aniline prior to the polymerization reaction; nitrogen sorption showed a decrease in pore volume after polymerization but not a complete disappearance of porosity. Electrical conductivity measurements showed only low levels of conductivity, a percolation network did not exist. However, the tubular structure of the channels was maintained after insertion of PANI. *Choi et al.* [293] explored the electrorheological properties of PANI-(MCM-41) nanocomposites. Since MCM displayed dielectric properties, a possible application of a colloidal suspension or dispersion of such particles showed a dramatic, increase in the effective viscosity upon application of electrical field. It was found, that the polarizability of the nanocomposites is much larger than those of the pure MCM material. In a subsequent study [294], structural investigations hint to a partial filling of the channels of the MCM material Figure 63. Long conductive wires were formed resulting in an electrically anisotropic material. PANI-(MCM-41) based ER fluids showed a much larger stress increase under applied electric fields than other comparable materials and than PANI, maintaining a high stress level even up to shear rates above 100 s^{-1}.

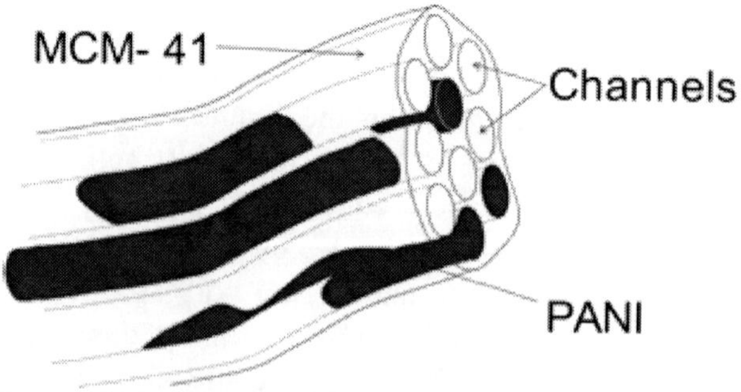

Reproduced with permission from ACS "Enhanced Electrorheology of Conducting Polyaniline Confined in MCM-41 Channels"M.S. Cho, H.J. Choi, W.S. Ahn, Langmuir ,2004 ,20(1),202-210.

Figure 63. PANI-(MCM-41) composite particle.

PANI-(MCM-41) composites display also semiconductor behaviour besides paramagnetic characteristics as shown by *Anunziata et al.* [295]. *Densakulprasert et al.* [296] studied the effect of zeolite content, pore size and ion exchange capacity on electrical conductivity response to carbon monoxide (CO) of polyaniline/zeolite composites. Zeolite Y, 13X, and synthesized Al-MCM41, all having the common cation Cu^{2+}, were dry mixed with synthesized maleic acid (MA) doped polyaniline and compressed to form polyaniline PANI / zeolite pellet composites. The Y, 13X and AlMCM41 zeolite have nominal pore sizes of 7, 10 and 36 Å respectively. With an addition of 13X zeolite to pristine polyaniline, the electrical conductivity sensitivity to CO/N_2 gas increases with zeolite content. For the effect of zeolite type, the highest electrical conductivity sensitivity is obtained with the 13X zeolite, followed by the Y zeolite, and the Al-(MCM-41) zeolite, respectively. The poor sensitivity of zeolite Al-(MCM-41) was probably due to its very large pore size and the lowest Cu^{2+} exchange capacity compared to the other two porous materials. Y zeolite and 13X zeolite have comparable pore sizes but the latter has a greater pore free volume and a more favourable location distribution of the Cu^{2+} ions within the pore. The temporal response time increases with the amount of zeolite in the composites but it is inversely related to the amount of ion exchange capacity. The PANI-(SBA-15) nanocomposites exhibited similar properties as the MCM nanocomposites, although the host pores were larger [297]. However, in this case the electrorheological properties of the PANI-(SBA-15) composite suspensions were lower than those of pristine PANI

6.4. PANI- CNT NANOCOMPOSITES

When used as fillers in polymeric matrices the high aspect ratio of CNT results in the formation of a connected CNT-network at low volume fraction. The transition is known as the percolation threshold. A marked increase in the mechanical strength and electrical conductivity are observed above percolation, connected network simultaneously provides a mechanical backbone and a pathway for electrical conductivity. Electrical percolation thresholds below 0.1 wt % of SWNT were reported in several polymeric matrices. However, although the reported percolation values are about two orders of magnitude lower than the value for carbon black, these values are many times higher than the theoretical predictions for randomly oriented objects of similar aspect ratio. The measured values were often attributed to experimental flaws such as a low quality of the electrical contacts between polymer coated CNT and originated from intermolecular interactions. *Szleifer and Yerushalmi-Rozen* [298] reviewed the behaviour of mixtures of polymers with carbon nanotubes with a detailed discussion of the use of polymers as dispersing agents of individual CNT. There several reports on the combination of PANI with CNTs in the literature. The first report goes back to 1999 (*J.N.Coleman et al.* [299], a time shortly after the discovery of the CNTs. Although this report did not deal with PANI as matrix material but with a derivative of PPV, it was already discovered that doping of such a polymer with conductive CNTs led to an increase in conductivity by ten orders of magnitude at percolation thresholds of about 8.5 %! An intimate contact between the matrix (PANI) and the CNTs was realized by an in-situ synthesis of the PANI-CNT composite as described firstly by *Maser et al.* [300-301]. Here the CNTs (MWCNT) were dispersed in diluted HCl under sonication. In a subsequent step aniline and ammonium persulfate [APS] were added, the polymerization was performed at 0-5 °C for 24 hrs. The resulting hybrid product displayed a conductivity of about 0.5 S/cm for a load of 20 % CNTs. *Deng et al.* [302] described an in-situ polymerization of the PANI-CNT hybrids with a subsequent doping using H_2SO_4 resulting in a 10 % composite with a conductivity of 0.66 S/cm.In a subsequent paper PANI-CNT hybrid was prepared by a different route [303]. Here, water, divinylbenzene and DBSA were mixed with the CNTs, aniline and APS were added and polymerized at 0-5 °C for 12 hrs resulting in a hybrid material with a conductivity of 0.066 S/cm. *Zengin et al.* [304] describes a similar in-situ synthesis of hybrid materials. This procedure led to a fine dispersion of the CNTs and peeling off of bundles into individual tubes. A coating of PANI onto the CNTs was described by Li et al. [305]; 8 % CNTs in the composite resulted in a 14 fold increase of conductivity of the PANI i.e. 1.26 S/cm. Electrochemical polymerization procedures using similar reactants and mixtures resulted in slightly larger conductivity values of 1.56 S/cm for 8 % CNT in PANI [306].The CNTs were reported to disperse readily in aniline; therefore the CNTs were refluxed in the dark for 5 h in aniline previous to the in-situ synthesis [306-307]. After evaporation of the solvent, the main agglomerate was broken up and loosely entangled CNTs were obtained. Alternatively, surfactants were used to facilitate the dispersion of CNTs previous to the in-situ synthesis of PANI, Figure 64,65 [309]. Here, the dispersion of CNTs is realized in a water CTAB mixture followed by the PANI synthesis. With increasing amount of aniline, the thickness of the resulting coating on the CNTs increased.

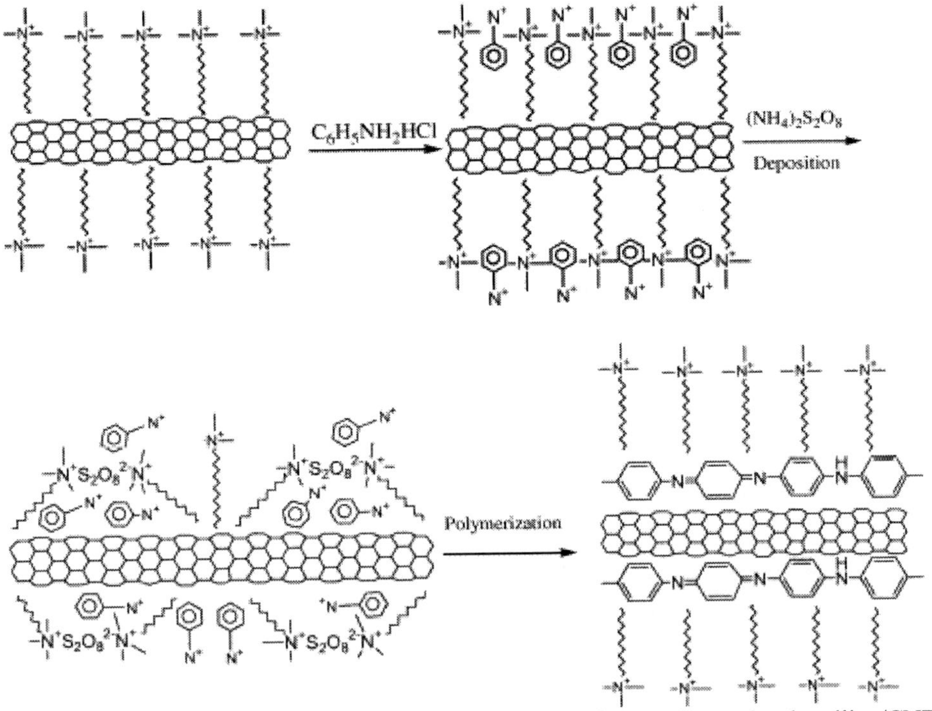

Reproduced with permission from Elsevier "Cationic surfactant directed polyaniline/CNT nanocables: synthesis, characterization, and enhanced electrical properties" X Zhang , J Zhang , R Wang , Z Liu Carbon 2004,42,1455-1461.

Figure 64. Dispersion of CNTs using CTAB and subsequent coating via in-situ synthesis of PANI.

A hybrid containing 15 % CNTs displayed a 100 fold increase in the conductivity (ca 1 S/cm) compared to the values of pure PANI and a 25 % CNT containing hybrid displayed a conductivity of 1.35 S/cm. Another surfactant (Triton X-100) was used for achieving better dispersions of CNTs previous to the in-situ synthesis of PANI-CNT acting as seed for PANI nanofibres.The method resulted in a hybrid containing ca. 1 % MWCNTs with a conductivity level of 3.9 S/cm, [310].Yet another synthetic strategy to obtain hybrid PANI-CNT materials or CNTs coated with PANI was described by W*u et al.* ,Figure 66[311]. They use of an inverse microemulsion polymerization of PANI in butanol/hexane stabilized by SDBS. The CNTs were added to the microemulsion and sonicated until a homogeneous solution was obtained. Aniline and APS were added and polymerized; the resulting material showed a particulate coating of PANI on the nanotubes. Figure 67 .*Wei et al.* [312] described the use of sulfonated MWNTs, and *Zhang et al.* [313] and *Wu et al.* [314] ,Figure 67 the use of carboxylated CNTs.

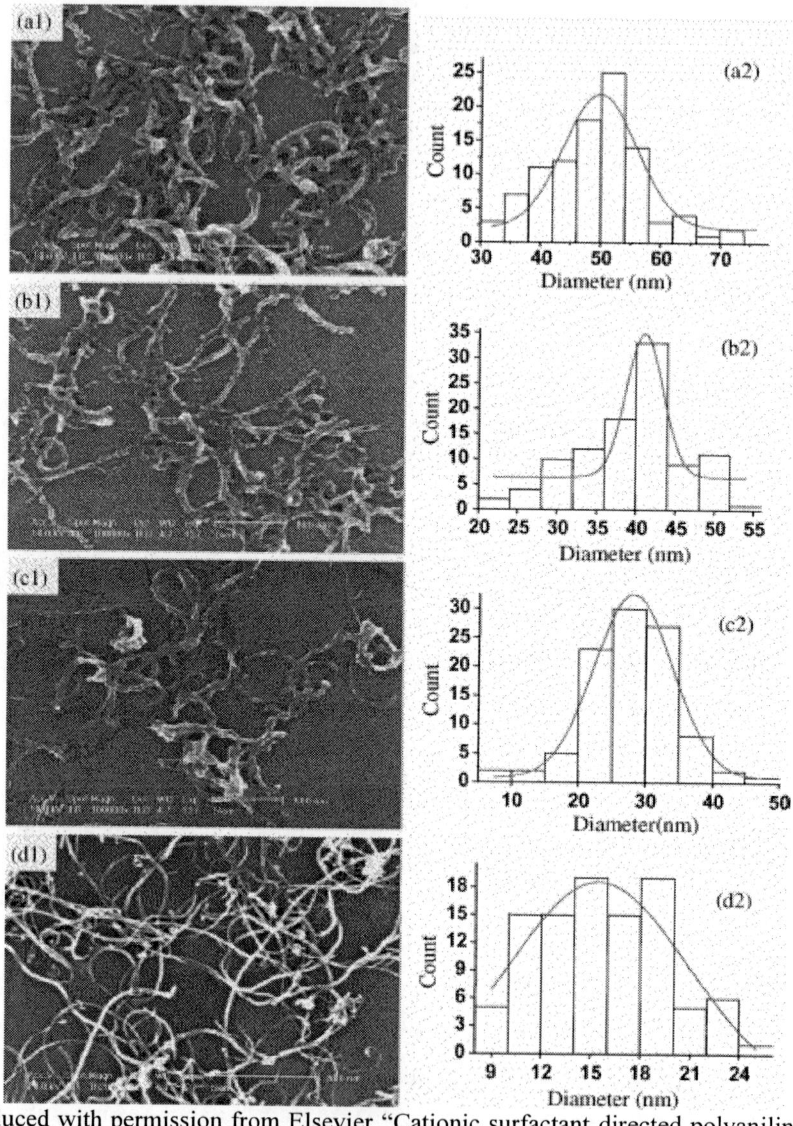

Reproduced with permission from Elsevier "Cationic surfactant directed polyaniline/CNT nanocables: synthesis, characterization, and enhanced electrical properties" X Zhang , J Zhang , R Wang , Z Liu Carbon 2004,42,1455-1461.

Figure 65. SEM pictures of CNTs wrapped by PANI using different ratios and distribution of CNT.

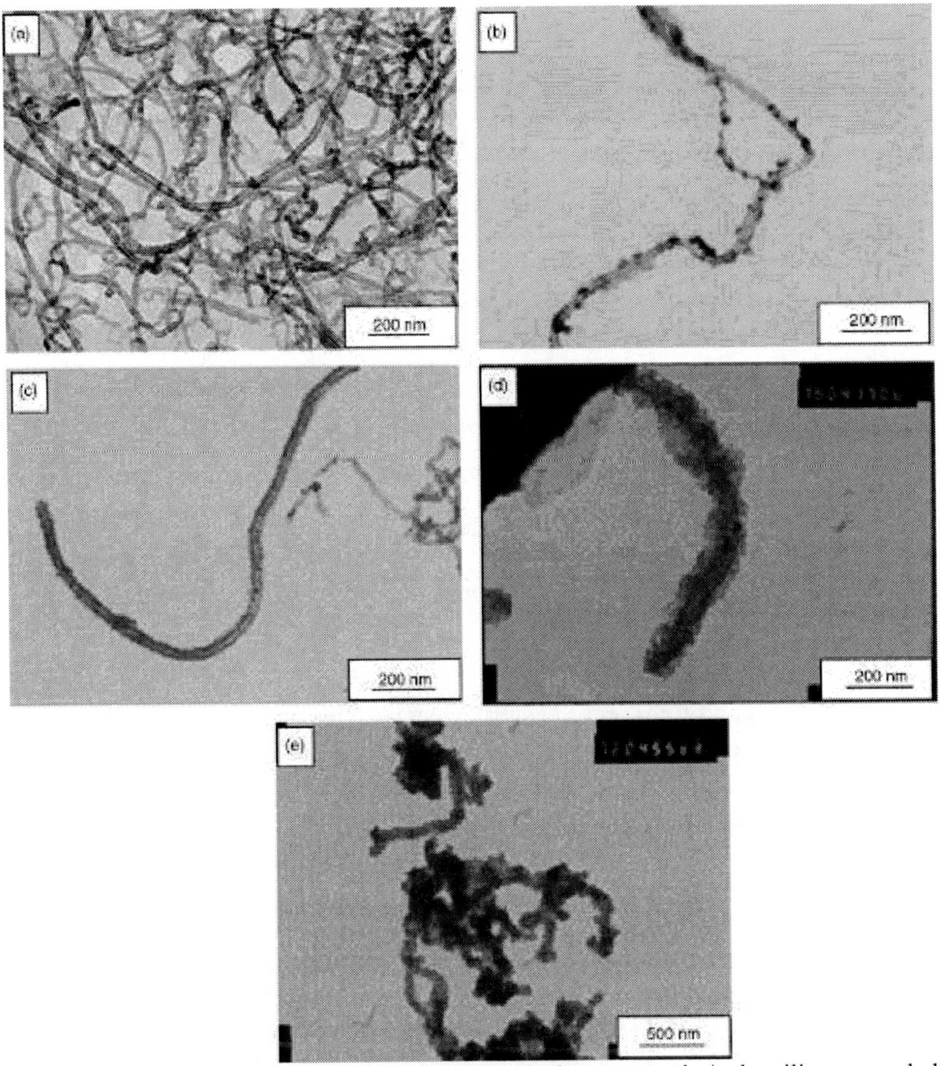

Reproduced with permission from Elsevier "Carbon nanotube/polyaniline core-shell nanowires prepared by in situ inverse microemulsion", Y.Yu, B. Che, Z. Si, L. Li, W. Chen, G. Xue ,Syn Met, 2005,150,271-277.

Figure 66. PANI-CNT materials using inverse microemulsion polymerization.

Although the coating of PANI onto the sulfonated CNTs was rather homogeneous, the increase in conductivity was only moderate (0.02 S/cm). In case of the carboxylated CNTs also homogeneous coatings of PANI were obtained with a conductivity value of 52 S/cm for hybrids containing 1 wt % MWCNTs.CNTs can also be functionalized to act as monomeric partner in the PANI polymerization process.

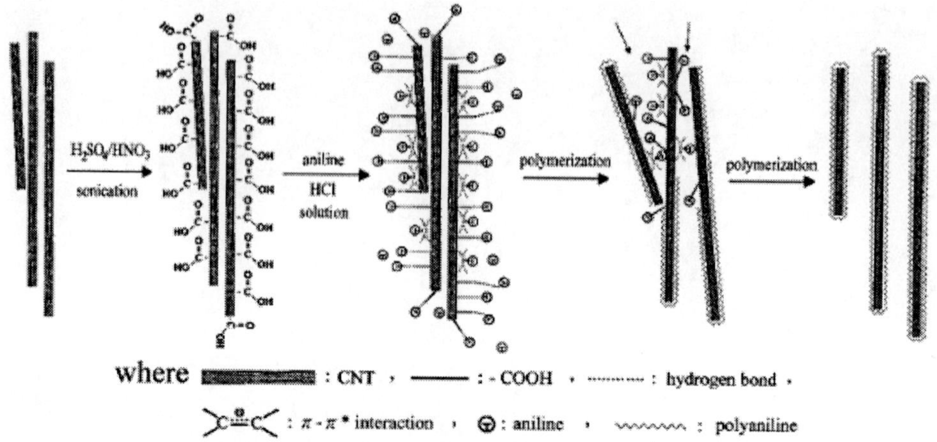

Synthesis of hybrid PANi-CNT materials using inverse microemulsion polymerization and resulting coating onto the tubes after Wu et al. [218].
T-M Wu, Y-W Lin, C-S Wu, C-S Liao ,Carbon ,2005,43,734-740.

Figure 67. Synthesis of hybrid PANI-CNT materials using inverse microemulsion polymerization.

Philip et al. [315-316] discussed the transfer of oxidized CNTs using thionyl chloride into acid chlorides and the subsequent reaction with p-phenylenediamine to amine functionalized CNTs which can act as reaction partners with aniline to form a covalently attached PANI coating onto the CNTs. A formation of hybrid materials containing PANI and CNTs via a common solvent is only reported in a few papers. *Baibarac et al.* [317] and *Sainz et al.* [318] used NMP as common solvent and obtained hybrid materials with conductivities of 0.1-0.3 S/cm (50% MWNT). Also the use of DMF as common solvent is reported resulting in an increase of conductivity up to 2.5 orders of magnitude when used in polymer blend compared to the PANI / polymer blend without the use of CNTs. A direct chemical functionalization of CNTs with the aim to get to better dispersion properties may also directly influence their conductive properties as described recently by Park et al. [319]. Robust differences are found between monovalent and divalent functionalizations, for the former a small number of substituents can significantly disrupt the ballistic conductance of nanotubes near the Fermi level. As the concentration increases, the conductance decreases rapidly and approaches zero. In contrast, divalent functionalizations had weak effect, and the nanotube quantum conductance remained above 50% of that of a perfect tube even for a substituent concentration as large as 25%. These differences can be attributed to the formation of impurity states near the Fermi level for monovalent additions, while divalent addends create impurity states far away from the Fermi level.

In summary, CNTs can only be homogeneously and stably dispersed in polymeric matrices and as such also in PANI if the strong interaction between the CNTs themselves can be overcome by an intense interaction with the matrix polymer. Functionalization is only one option. So far the preferential option is the direct attachment of polymeric chains onto the CNT surface. Polyaniline–fullerene (PA-C_{60}) composites, obtained by mixing solutions of polyaniline–emeraldine base [PA-EB] and C_{60}, as well as by chemical synthesis from aniline, sulphuric acid and potassium dichromate, with addition of C_{60} show also evidence for interface inter cations as studied by *Giusca et al.* [320].

6.4.1. PANI – Graphite Nanocomposites Including Carbon Black

A facile process for the synthesis of exfoliated PANI / graphite nanocomposite via the microwave irradiation and sonication of synthesized expandable graphite followed by in-situ polymerization of aniline monomer in the presence of graphite nanosheets has been reported by *Du et al.* [321-322], Figure 68,Figure 69. The nanoscale dispersion of graphite sheets was evidenced by the SEM and TEM examinations. Electrical conductivity measurements indicated that the conductivity of final PANI / graphite nanocomposites increased greatly compared with pure PANI. The electrical conductivity of the PANI / graphite nanocomposite increased with increasing graphite content. The conductivity reached 33.3 S/cm with only 1.5 wt % graphite; this was 5 times higher than for pure PANI (~ 5 S/cm). The electrical conductivity of the nanocomposites levelled off with further increasing graphite content. This means that the percolation threshold in the conductivity of the nanocomposites was lower than 1.5 wt % and can be attributed to two facts: (1) the nanoscale dispersion of graphite nanosheets in the polymer matrix and the formation of the conducting network and (2) the interaction between the large conjugated structure of the graphite nanosheets and the quinoid ring of PANI. The graphite nanosheets within the PANI matrix serve as the electrically conductive bridges.

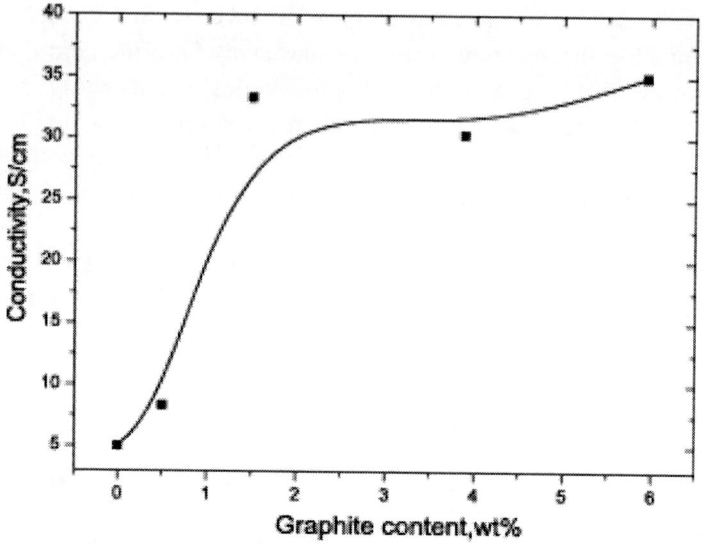

Reproduced with permission from Elsevier "Facile synthesis of highly conductive polyaniline/graphite nanocomposites" X.S. Du, M. Xiao, Y.Z. Mang, Eur Polym J, 2004,40,1489-1493.

Figure 68. Variation of volume electrical conductivity of PANI / graphite nanocomposite.

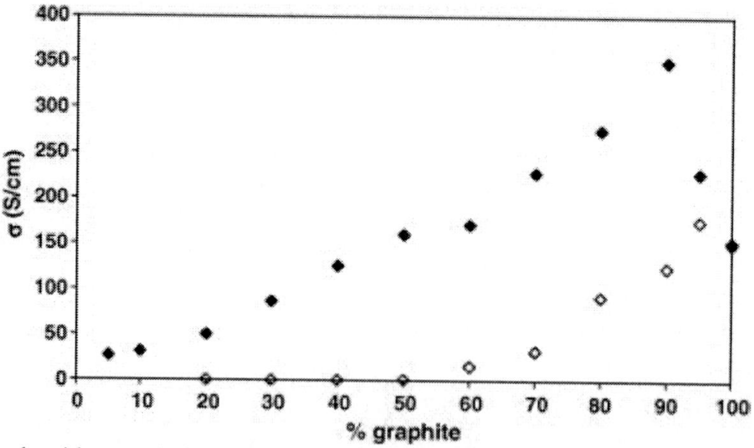

Reproduced with permission from Elsevier "Facile synthesis of highly conductive polyaniline/graphite nanocomposites" X.S. Du, M. Xiao, Y.Z. Mang, Eur Polym J, 2004,40,1489-1493.

Figure 69. Conductivity study of PANI / graphite composites: (♦) doped GP, (◊) dedoped GP.

A subsequent study by *Bourdo and Viswanathan* [323] extended the investigations to ratios of PANI / graphite between 0 and 100 % graphite. At a concentration of about 90 % graphite, a maximum in conductivity (350 S/cm) was reached exceeding the conductivity levels of both components (150 S/cm and 15 S/cm respectively),Figure 69. The higher conductivity of the composite was attributed to the polymer providing a matrix for the colloidal graphite to align, or vice versa, in such a way that the transfer of electrons between PANI and graphite becomes less restricted. The observation of two significant increases in conductivity suggests that two percolations may exist in the system: graphite into the PANI matrix and PANI into the graphite matrix. The observed changes with doping and dedoping indicate that the ICP played a significant role in the mechanism of conduction in these composite materials.

An increase in electrical conductivity of only 38.5 % has been reported for the case of carbon black as nanoparticle incorporated within PANI [324]. The core–shell nanocomposites of polyaniline and carbon black (PANI-CB) were electrochemically synthesized in acidic aniline and carbon black suspension. SEM and TEM observations reveal that the PANI synthesized in $HClO_4$ media exhibits one-dimensional growth and results in a porous PANI deposit constructed by coralloid interlinking branches having about 2 μm average diameter. In contrast, the existence of CB suspension produces rod-like PANI-CB core–shell nanocomposites, and decreases the average diameter to nearly 200 nm. XRD and Raman spectroscopy were employed to characterize the as-obtained CB powders, PANI and PANI-CB nanocomposites. Measurement of electric conductivity shows that incorporation on CB particles during electrochemical synthesis increased the conductance. PANI / carbon nanocomposite are being applied or discussed for application in many fields spanning from electromagnetic shielding and microwave absorption over actuators, construction of super capacitors, as battery materials and in fuel cells, as parts (electrodes) of pH, gas and many other sensors, for anti-corrosion protection, as electromagnetic fluids and many more areas.

7. Preparation Techniques

7.1. Electro Spinning Process

Electrostatic spinning or "electrospinning" is a process capable of producing polymer fibers in the nanometer diameter range, Figure 70. It is a novel and efficient fabrication process that can be utilized to assemble fibrous polymer mats composed of fiber diameters ranging from several microns down to lower than 100 nm.The basic setup for electrospinning consists of three major components: a high-voltage power supply, a spinneret (a metallic needle) and a collector (a grounded conductor) Figure 70. The spinneret is connected to a syringe in which the polymer solution (or melt) is hosted. With the use of a pump, the solution can be forced through the spinneret at a constant and controllable rate. When a high voltage is applied, the pendent drop of polymer solution at the nozzle of the spinneret will become highly electrified and the induced charges are evenly distributed over the surface. Consequently the drop will experience two major types of electrostatic forces: electrostatic repulsion between the surface charges and coulombic force exerted by the external electric field. Under the action of these electrostatic interactions, the liquid drop will be distorted into a conical object. When the strength of the electric field has surpassed a threshold value, the electrostatic forces can overcome the surface tension of the polymer solution and thus forces the ejection of a liquid jet from the nozzle. As the liquid jet is continuously elongated and the solvent is evaporated, its diameter can be greatly reduced from hundreds of micrometers to as small as a few nanometers. Attracted by the grounded collector placed under the spinneret, the charged fiber is often deposited as a randomly oriented, non woven mat.

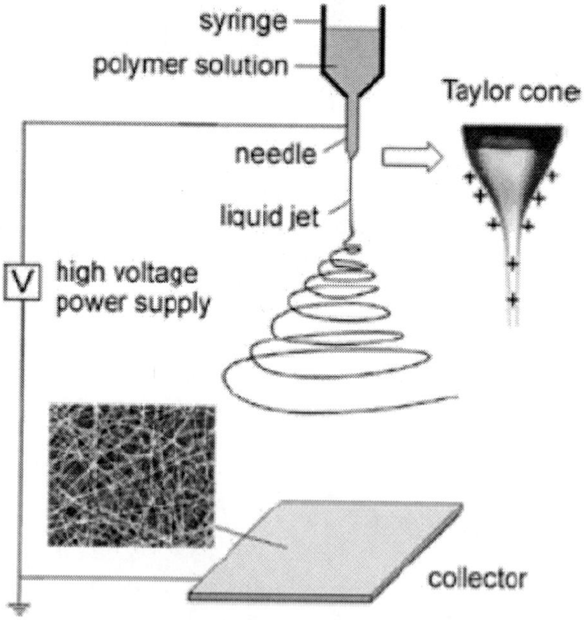

Figure 70. Schematics of electro spinning process.

The setup for electrospinning is extremely simple, but the spinning mechanism is rather complicated. Recent experimental observations demonstrate that the thinning of a jet during electro spinning is mainly caused by the bending instability associated with the electrified jet [325]. Theoretical studies clearly showed that the spinning process only involves whipping of a liquid jet. The whipping instability is mainly caused by electrostatic interactions between the external electric field and the surface charges on the jet. The formation of fibers of tiny diameters is mainly achieved by the stretching and acceleration of the fluid filament in the instability region [326].In general, electro spinning is a process with low productivity because the polymer solution has to be fed at relatively slow rates in order to obtain ultra thin fibers.

The following system and process parameters have been known to affect fiber diameter

- Viscosity
- Concentration
- Net charge density (conductivity)
- Surface tension of the polymer fluid.
- Molecular weight
- Molecular weight distribution
- Topology (branched, linear etc.) of the polymer.
- Electric potential
- Flow rate of the polymer solution
- Distance between the capillary-end and target/collection screen
- Ambient parameters (temperature, humidity and air velocity in the chamber)
- Motion of target screen
- Internal diameter of the nozzle/capillary

7.2. Mechano-Chemical Synthesis

In view of the technological difficulties of conventional methods (multi-stage character, long process duration, high synthesis temperature)the need for replacing ecologically hazardous production works that are usually accompanied by large amount of waste water and other wastes; mechano-chemical synthesis may be alternative as a dry, waste-free, ecologically safe method. Mechano-chemical approach to the synthesis of nanocomposites is likely to be the most promising one because it allows obtaining nanocomposites with various classes of compounds: metals, oxides, salts, organic compounds in various combinations. The mechano-chemical approach to the synthesis of substances has been successfully applied to different classes of chemical compounds: intermetallides and solid solutions, oxides and complex oxides, hydrides, organic compounds, etc. It is a promising method of obtaining not only nano-sized states but also non equilibrium phases [327-328] .When choosing components to obtain mechano-composites for a specific practical application, two factors should be taken into account. First, no chemical interaction should occur between these components, though, at the same time, the surface interaction should exist, in order to provide sufficient contact surface. For example, incase of metals, a good wettability should exist. Second, the stabilizing component of a mechano-composite should be inactive toward the external reagent. The use of mechano-composites can be considered as a solution to the problem of creating extremely large surfaces for active metals stable with respect to the oxygen of the air. For example, if non-interacting metals are taken as initial components, morphologically metastable structure with high density of inter-phase boundaries can be created mechanochemically. In such a system, nano-sized state of an active metal will be stabilized by inter-phase boundaries, while atomic contact between these metals excludes the oxygen of the air. One of the confirmations of the stability of these nano-sized states toward oxidation in the air is the fact that the specific surface of mechano-composites determined from gas adsorption is usually 1–2m2/g, while for nanometer-sized particles it should be one or two orders of magnitude larger. These metal mechano-composites with large surface of the active component can be used for the creation of metal cements, as a material for hydrogen storage, for preparation of Raney catalysts, etc. It is known that copper, nickel, and other metal powders are used to make diffusion-hardening metal glues, metal cements, conductive layers. These materials are usually obtained by mixing ultrafine metal powders with liquid metal eutectics. A new intermetallic compound is formed at the boundary between phases. The formation rate of this compound will further be determined by the diffusion of reagents through the layer of products. Since this process is controlled by diffusion, the interaction rate effectively depends on particle size. The use of nano-sized metals allows substantial improvement of the consumer properties of materials. However, their instability towards the oxygen of the air brings about technological difficulties into their usage and into the quality of the resulting soldering material. The use of mechano-composites with nanometer-sized active component for these purposes allows one to avoid the problem connected with the effect of oxygen of the air on the process. The systems Cu–Bi and Fe–Bi with low bismuth content can be considered as one of such 'immiscible' systems for obtaining nanocomposites. Investigations into the formation of such mechano-composite structures have been performed. It is known that no chemical interaction occurs between the components in the systems Cu–Bi and Fe–Bi, but at the same time bismuth wets well the surfaces of both copper and iron. Hence, mechano-composites are possible in these systems. For bismuth content below 3 at %, the investigation of MA process in the systems Cu–Bi and Fe–Bi demonstrated a decrease in the intensity of basic reflections of bismuth with time of mechano-chemical treatment without any shift of copper reflections. Under the activation conditions of our experiments, the reflections of bismuth disappear almost completely within 10 min. The diffraction reflections of copper broaden remarkably without changing their position. Composites of natural layered silicates (kaolinite, talc, pyrophyllite, etc.) with various organic compounds (saturated, unsaturated, hydroxo acids, fatty alcohols, etc.) have been used to prepare materials for the solid-phase curative cosmetics and biologically active nutrition additives A biologically active organic constituent is distributed homogeneously over the silicate surface and can be easily extracted. At the same time, neutralization of the protons of an acid provides the absence of irritating action of acids. The activation of layered silicates in the presence of solid-phase organic acids is also accompanied by structural changes but without dramatic consequences. The changes that occur in kaolinite when treated with stearic acid after activation for 3 min provide the evidence that the structure of kaolinite is conserved, though hydrogen bonds between the layers of nets are broken. The X-ray data also confirm the conservation of the crystal structure of the silicate. The lattice parameter remains unchanged, which also suggests that organic acids do not enter the interlayer space. It is most reasonable to assume that during silicate grinding in course of mechanical activation, the basic centers on the newly opened planes (surfaces) are neutralized by the protons of acids, i.e. mechano-chemical bonding of organic compounds to the silicate surface is formed. The absence of diffraction reflections of organic acid is the evidence of almost complete destruction of its crystal structure. Electron microscopic investigations of these composites show that nano-size particles may be obtain by mechanical activation of layered silicates in the presence of organic acids without the destruction of their crystals structure On one hand, mechano-composites allow achieving ultrafine size for organic acids; on the other hand, they facilitate grinding of layered silicates without destroying their crystal structure. A short-term (up to 3 min) mechanical activation of a mixture of two components in the high-energy ball mill is the proficient method of creation of various nanocomposites. These mechano-composites, on one hand, as systems with unusually large contact surface can serve as precursors for subsequent chemical interaction of their components in mechano-chemical, thermal, and other processes in which not only the synthesis temperature is dramatically decreased, but also the phase composition and morphology of the products are substantially changed. By changing the parameters of mechano-composites one can affect all the above-listed processes. On the other hand, a mechano-composite stabilize extremely large surface of active components. It is difficult to obtain and conserve them by other methods. This is a real route to create materials with new properties: solid-phase curative cosmetics with decorative properties, pigments, new forms of pharmaceuticals, materials for hydrogen storage, metal cements, precursors for Raney catalysts, etc.

7.3. Vapor-Phase Assisted Surface Polymerization (VASP)

This technique has been developed as the simplest method for constructing micro architectures on solid substrate surfaces with the advantages of being solvent less and precise. The VASP technique, Figure 71 has other advantages making it useful for the construction of fine structured composites [329] and coatings [330] at the gaps and spaces within the solid substrate. When these gaps and spaces are wider than the size of monomer molecule, the vaporized monomer can diffuse and penetrate interstitially within the solid substrates.

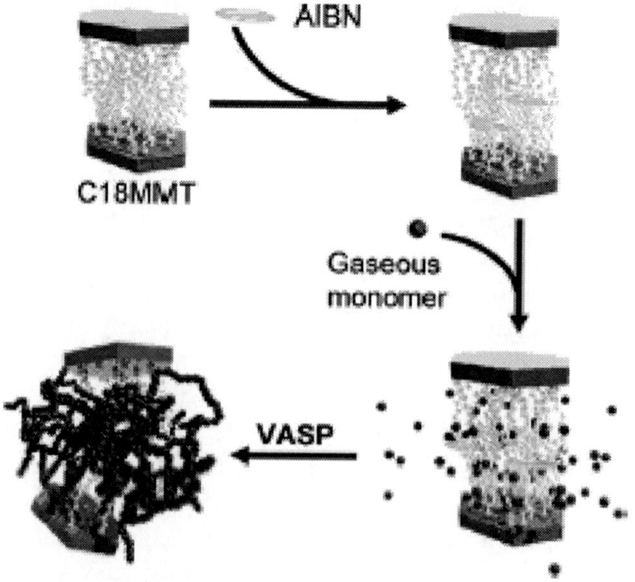

Figure 71. Exfoliation of silicate layers by VASP.

After the diffusion and adsorption on the interstitial surfaces, the monomers polymerize in a manner of "pseudo-grafting "at the substrate surfaces. Polymer chains then grow on the surfaces by filling the spaces, giving a superior anchoring effect, with excellent binding strength at the polymer/ substrate interface. This technique has great potential in the construction of nanocomposites by combinations with inorganic materials having nanoscale structures and can proceed in the simplest manner without any solvent [331].

7.4. Intercalation Techniques

There are several processes to make clay-based polymer nanocomposites, including *in-situ* polymerization, solution exfoliation and melt intercalation. As shown in Figure 72, each technique consists of several steps to achieve polymer nanocomposites and begins with organoclays or sometimes pristine clays. In the *in-situ* polymerization, monomers are intercalated into layered clays and subsequently polymerized within the gallery via heat, radiation, pre-intercalated initiators or catalysts.In the solution exfoliation, layered clays are exfoliated into single platelets using a solvent in which the polymer is soluble. The polymer is then mixed with the clay suspension and adsorbed onto the platelets. The solvent is finally eliminated from the clay polymer complex through evaporation. In the melt intercalation, layered clays are directly mixed with the polymer matrix in the molten state

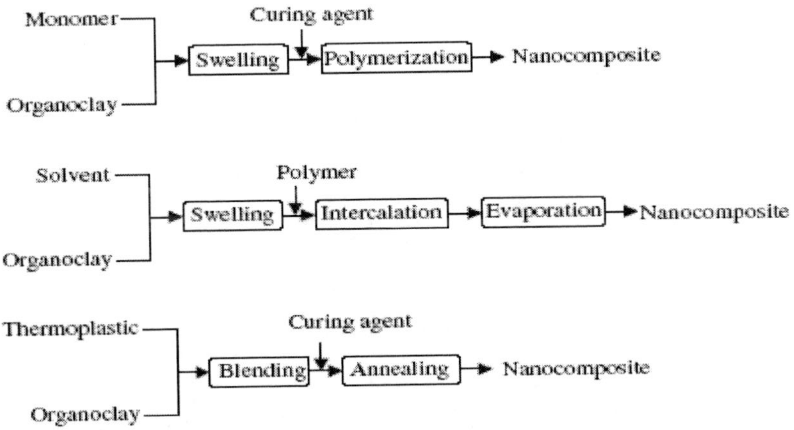

Figure 72. Flowchart of three processing techniques for clay-based polymer nanocomposites: *in-situ* polymerization (upper), solution exfoliation (middle) and melt intercalation (bottom).

When layered clays are filled into a polymer matrix, either conventional composite or nanocomposite ,Figure 73, can be formed depending on the nature of the components and processing conditions. Conventional composite is obtained if the polymer can not intercalate into the galleries of clay minerals. There are two extreme nanostructures resulting from the mixing of clay minerals and a polymer providing favorable conditions. One is intercalated nanocomposite (I), in which monolayer of extended polymer chains is inserted into the gallery of clay minerals resulting in a well ordered multilayer morphology stacking alternately polymer layers and clay platelets and a repeating distance of a few nanometers.The other is exfoliated or delaminated nanocomposite (II), in which the clay platelets are completely and uniformly dispersed in a continuous polymer matrix. However, it should be noted that in most cases the cluster (so-called partially exfoliated) nanocomposite (III) is common in polymer nanocomposites.

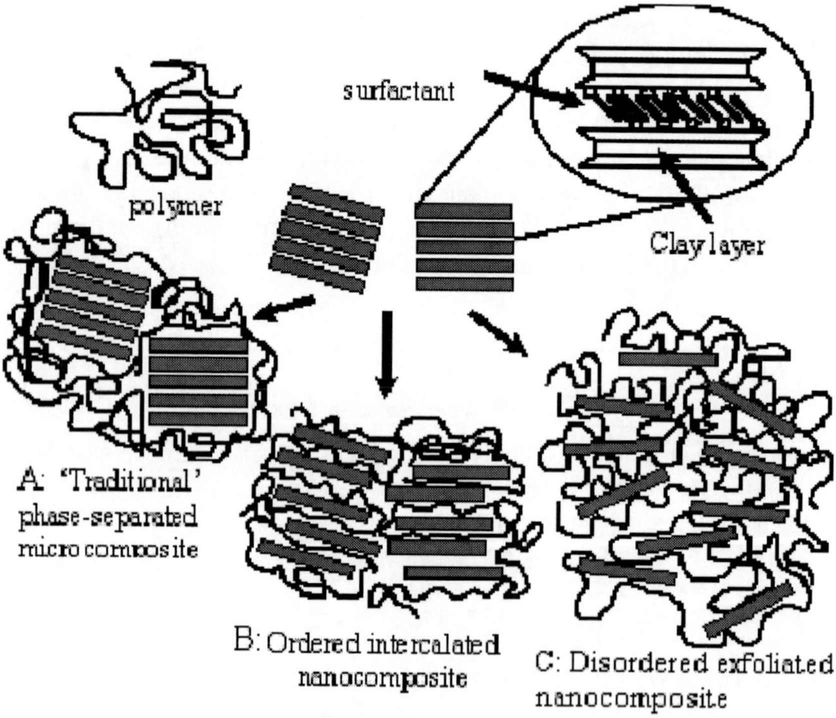

Figure 73. Schemes of clay-based polymer composites, including conventional composite and nanocomposite with intercalated (I), exfoliated (II) or cluster (III) structure.

Table 5. Common characterization techniques for clay based polymer nanocomposites

Techniques	Characteristics and properties
XRD/WAXRD	Degree of swelling and interlayer distance of clays Dispersion degree of clay platelets Morphology (conventional, intercalated, exfoliated or mixed) Kinetics of intercalation process
SEM	Surface roughness and morphology Dispersion degree of clay particles
TEM/HRTEM	Morphology and its development Microstructure (intercalated vs exfoliated) Spatial distribution of clay platelets Structural heterogeneities Defect structure and atomic arrangement
AFM	Crystallization behavior of polymer Surface roughness Particle size and distribution Morphology and microstructure
FTIR	Component identification and analysis Interfacial interactions Crystallization and orientation of polymer
NMR	Local dynamics of polymer chains Morphology and dispersion of clay particles Surface chemistry
SAXS	Dispersion of nanoscale clay platelets Morphology (intercalated, exfoliated or mixed) and its development Phase behavior and structure evolution Lamellar texture and thickness
TGA DSC	Thermal stability Melting and crystallization behavior Local dynamics of polymer chains
Cone calorimetry	Flame retardancy, such as heat release rate and carbon monoxide yield Thermal stability
Rheometry Mechanical test	Nanorheology Young's modulus Tensile strength Elongation at break Viscoelastic properties

8. Characterization Techniques

Various techniques have been employed to study polymer nanocomposites as summarized in Table.5.The structure of polymer nanocomposites is generally investigated by X-ray diffraction (XRD) and transmission electron microscope (TEM).XRD, in particular wide angle XRD, is the most commonly used technique for examining the structure and occasionally for studying the process kinetics of polymer nanocomposites.Either intercalated or exfoliated nanostructure can be studied by monitoring the position, shape, and intensity of the basal reflections of XRD patterns of the materials. For instance, the extensive layer separation in an exfoliated nanocomposite is reflected in the disappearance of any coherent XRD peaks whereas the finite layer expansion in an intercalated nanocomposite is associated with the appearance of a new basal reflection corresponding to the larger gallery height.XRD offers a convenient method to determine the interlayer spacing of original layered clays and intercalated nanocomposites. Unfortunately, such technique cannot provide information about the spatial distribution of the clay layers or structural non-homogeneities in nanocomposites.It is also difficult to study systematically systems showing a broadening peak and weakening intensity. From this point of view, information from XRD patterns is not sufficient to reveal the formation mechanism and ultimate structure of nanocomposites.

In contrast, TEM can provide direct qualitative information of structure, morphology and spatial distribution of the various components as well as the defect structure. However, one should be cautious about data from XRD when the layer spacing in intercalated nanocomposites exceeds 6–7 nm or when the layers become relatively disordered in exfoliated nanocomposites.In this case, the simultaneous use of small angle X-ray scattering (SAXS) with wide angle XRD can yield quantitative characterization of the structure in polymer nanocomposites.Besides, nuclear magnetic resonance (NMR) is another important tool for probing surface chemistry and coordination in exfoliated polymer nanocomposites, which may help quantify the level of clay exfoliation. Fourier transform infrared (FTIR) and Raman spectroscopy can be used to understand the structural formation of polymer nanocomposites.

9. Properties of Nanocomposites

9.1. Mechanical Properties

The mechanical behavior of nanocomposites is based on the nature of the polymer matrix, particularly crystalline or amorphous nature of the polymer, and the interaction between the filler and matrix. An increase in yield and tensile strength and modulus in nanocomposite systems as compared to microcomposites can be partially explained on the basis of the interaction between the filler and the matrix. It has been found that a greater adhesion between the matrix and inclusion causes less debonding when a stress is applied and, consequently the elastic modulus and strength are improved [332].The elastic modulus tends to increase with the volume fraction of inclusions in every case. In some systems, there is a critical volume fraction at which aggregation occurs and the modulus goes down [333-336]. In general, there is also an increase in modulus as the size of the particle decreases. Interaction between matrix and filler may play an important role in the effects of the nanoparticles on composite properties. [334-337] For polymer systems capable of having a higher degree of crystallinity, the increase in modulus with decreasing particle size is found to be greater in systems with poor interaction between filler and matrix as opposed to those with good interaction. However, the overall trend of the modulus of polymer nanocomposites is not found to be greatly dependent upon the nature of the matrix nor the interaction between filler and matrix. An examination of the yield stress gives a different trend than that of the elastic modulus. For composites with good interaction between filler and matrix, the yield stress tends to increase with increasing volume fraction and decreasing particle size, similarly to the increase in modulus under same conditions. The pattern changes when there is poor interaction between the matrix and particles. The addition of nanoparticles with poor interaction with the matrix causes the yield stress to decrease, compared to the neat matrix, regardless of the filler concentration or size. The ultimate stress follows a similar pattern as that observed for the yield stress. It generally increases in polymer systems (both crystalline and amorphous) with good filler–matrix interaction and increases in general as particle size decreases. There is no uniform trend with respect to the volume fraction of particles for the ultimate stress. A poor filler–matrix interaction leads to a decrease in the ultimate and yield stress as compared to the pure matrix system, Figure 74.

The change in strain-to-failure behavior for nanocomposites, like the yield and ultimate stress, is different depending on the system. In this case, however, the change is with respect to the nature of the polymer matrix as opposed to the filler–matrix interaction. In general, the addition of nanoparticles to a mostly crystalline or semi-crystalline polymer, regardless of the filler–matrix interaction, reduces the maximum strain. The opposite trend is found in amorphous polymers with the increase in strain-to-failure coinciding with a decrease in particle size [336-337].

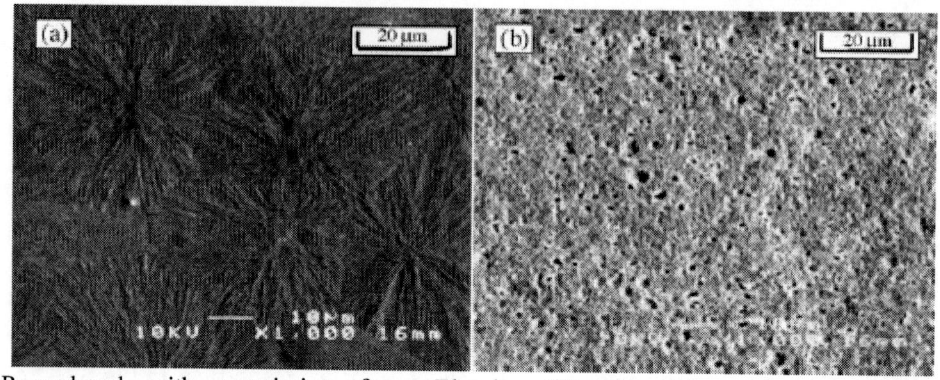

Reproduced with permission from Elsevier "Experimental trends in polymer nanocomposites—a review", J. Jordana, K.I. Jacob, R.Tannenbaum, M.A. S.I.Jasiuk* , Materials Science and Engineering A 393 (2005) 1–11.

Figure 74. (a) Pure polypropylene (b) polypropylene with filler.

9.2. Viscoelastic Properties

The small inter particle distance in nanocomposites is used as another parameter to explain the changes in the elastic modulus and strength of these materials when compared with the composites with micron-sized particles. In general, viscoelastic properties tend to be higher in nanocomposites than in pure polymer systems. When there is good filler–matrix interaction, the storage modulus generally increases with increasing volume fraction. The modulus also seems to increase as the particle size decreases. However, there is little experimental work in this area for composites with poor filler–matrix interactions.Overall, the storage modulus tends to increase with the presence of nanoparticles in a composite system. Morphological details, such as exfoliation, intercalation, or cross-linked matrix versus uncross-linked matrix, have a significant effect on the viscoelastic properties of nanocomposites. In epoxy–clay nanocomposite system forces from cross-linked epoxy molecules cause intercalated clay galleries to exfoliate, associated with a gradual increase in viscosity and a relatively fast increase in storage modulus [338-339]. In maleated polypropylene–layered silicate nanocomposites, exfoliated nanocomposite systems were reported to have higher shear and complex viscosities, while in a shear flow exhibiting the largest drop in complex viscosity due to alignment of clay layers [340]. Dynamic storage moduli also show a similar behavior [340]. Experiments with clay–nylon- 12 nanocomposite systems have shown that melt processing under low shear improves intercalation, while larger shear enhances exfoliation associated with a decrease in melt viscosity [341].

9.3. Crystallinity

When there is good filler–particle interaction, the glass transition temperature tends to increase with a decrease in the size of particles for amorphous polymers. For crystalline polymers, the transition temperature decreases with an increase in particle concentration. For an amorphous system with poor filler–polymer interfacial interaction, the glass transition temperature decreases overall. Thus, while the degree of crystallinity is not significantly affected by the presence of particles, the glass transition temperature is very dependent upon this factor.

9.4. Density/Volume Change

Investigations examining the change in the density or volume of polymer nanocomposites are limited. Nevertheless, the trend is that nano-sized inclusions affect the volume of a composite more than micron-sized inclusions do. The volume for the composite is higher for a nanocomposite than that of a micro composite with the same filler weight and the same polymer weight. It is important to point out that these results are only for composites with relatively good filler–polymer interaction.

However, there is no universal trend that can be modeled and explained. Some observations reveal that the behavior of nanocomposites is different from the composites with larger scale inclusions. The particle size particle morphology and the polymer tend to play a very important role. In addition, the nature of dispersion and aggregation of particles can affect the properties of composites significantly. Filler–matrix interaction is another factor that influences the properties. The strength of the interaction plays a role in the thickness and density of the interphase, which consists of a layer of high density polymer around the particle. The effect of the interface on the behavior of a composite depends upon the interparticle distance. For constant filler content, with reduction in particle size, number of filler particles increases, bringing the particles closer to one another. Thus, the interface layers from the adjacent particles overlap, altering the bulk properties significantly. These issues play a major role in the effect of nano-sized inclusions in a polymer matrix. For nanoparticles, any configuration changes in the matrix will have a significant effect when the characteristic radius of polymer chains is of the same order as the inclusions [341].

9.5. Thermal Properties

The thermal stability of polymer composites is generally estimated from the weight loss upon heating which results in the formation of volatile products.The improved thermal stability in polymer nanocomposites is due to the clay platelets which hinder the diffusion of volatiles and assist the formation of char after thermal decomposition. Another thermal behavior is the heat resistance upon external loading which can be measured from the heat distortion temperature(HDT).

9.6. Barrier Properties

Polymer nanocomposites have excellent barrier properties against gases (e.g., oxygen, nitrogen and carbon dioxide), water and hydrocarbons. Studies have showed that such reduction in permeability strongly depends on the aspect ratio of clay platelets, with high ratios dramatically enhancing gaseous barrier properties. The enhanced barrier properties of polymer nanocomposites may be explained by the labyrinth or tortuous pathway model,Figure 75.

When a film of polymer nanocomposites is formed, the sheet-like clay layers orient in parallel with the film surface. As a result, gas molecules have to take a long way around the impermeable clay layers in polymer nanocomposite than in pristine polymer matrix when they traverse an equivalent film thickness. It is interesting to note that the enhancement of barrier properties does not arise from the chemical interactions since it does not depend on the type of gas or liquid molecules.

9.7. Electrical Conductivity

Electrical conductivity of the nanocomposites have been found to depend upon the method of preparation of the nanocomposites. Nanocomposites with conjugated conducting polymers such as PANI, polypyrrole (PPy), and polythiophene (PTP). are prepared by three different ways - in-situ polymerization induced by interlayer transition metals (e.g., Cu^{2+}, Fe^{3+}) ; using external oxidizing agents (e.g., ammonium peroxodisulphate) and addition of polymer during the hydrothermal synthesis of clay minerals. By using the firs approach, PANI nanocomposites prepared from fluorohectorite and MMT exhibited an intercalated structure as demonstrated by XRD and FTIR analyses. Moreover, a significant increase in the electrical conductivity was observed, which attained high in-plane direct current (DC) conductivity (5×10^{-2} S/cm) with an anisotropic ratio close to [105-161] . Similar materials have also been formed from MMT by the second technique, which exhibited conductivity (10^{-6} S/cm) much lower than the former one. Such a difference is probably due to the nature of PANI chains obtained under various conditions and the anisotropy of material conductivity. *Mehrotra et al.* [162] also reported the formation of PPR nanocomposites with Cu^{2+-} fluorohectorite. Moreover, doping of these nanocomposites with iodine vapors produces a strong increase in the DC conductivity from $2\times^{10-5}$ S/cm to $1\text{-}2\times10^{-2}$ S/cm and a high anisotropic ratio (4×10^{3}) as well.

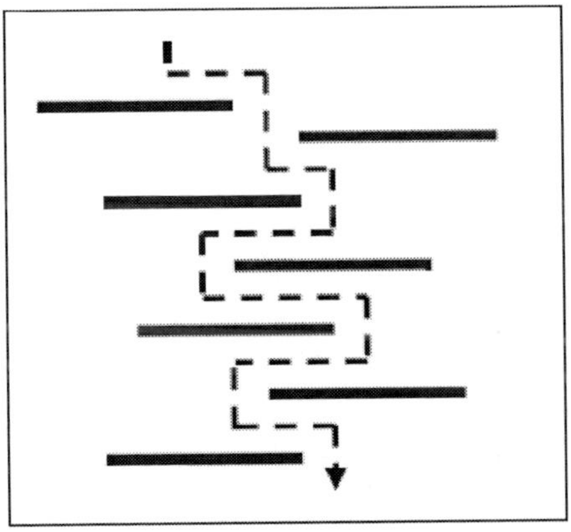

. Figure 75. Proposed model for the tortuous zigzag diffusion pathway of a gas through clay-based polymer nanocomposites

10. Design, Nano Fabrication and Nano Integration Application of Nano Conducting Polymers and Nancomposites

Design of nano-structured materials has recently become an outstanding combination of chemistry and materials science in applications. Developing such systems to obtain different nano-shapes, such as; tubules, rods, fibers, wires, arrays, nozzles etc. has been of great interest to so many futuristic scientists in each branch of science. A few examples to nano-structured production of materials are porous membrane template synthesis, zeolite template synthesis, layer-by-layer inorganic assembly, and so on. After the discovery of conjugated polymers in the late 1970s, the chemical route turned towards the nanofabrication of these materials to enhance intrinsic properties. Production of nano-shaped conducting polymers was first achieved by *Martin et. al.* using template synthesis method. One of the earliest applications of this method was exampled as the preparation of nanoscopic electrodes within the pores of commercially available polymeric membranes. Both chemical and electrochemical synthetic methods have been applied to the preparation of the template-synthesized polymers. Ina further advancement , the chemical growth of PPy nano-tubules and nano-fibrils was obtained from commercial polycarbonate (PC) microporous and nanoporous track-etched membranes (PTM)[7].In another study, organic PPy and poly(N-methylpyrrole) (PNMPy) microtubules were produced with electrochemical method using a microporous membrane template. More recently, PPy nanocylinders were fabricated chemically within the pores of PC-PTM [342]. Although most of the first studies have been reported on PPy, there are some other studies that report template-grown nano-structured poly (thiophene) (PTh). Nano-wires of poly (3,4- ethylene dioxythiophene) (PEDOT), a PTh derivative, have also been prepared by confined electrochemistry in nanoporous PC membranes. PEDOT's high environmental stability is an advantage for the template based synthesis. The polymerization of 3,4- ethylene dioxythiophene (EDOT) inside the pores of track-etched membranes was achieved successfully. In another perspective, template electrochemical synthesis of PTh was studied using different kinds of silica sol-gel as the host matrix [344].Nano-fibrilar copolymers of pyrrole and thiophene were achieved electrochemically inside the pores of microporous anodic aluminum oxide (AAO) template membranes. Similar methods were applied to the synthesis of unidirectional conducting polymer assemblies of PPy and PNMPy nano-particles in alumina membrane. The concept of nano-integration is shown in Figure 76.

Typical nanomaterials are atomic and molecular layers on nano-scale, π-conjugated polymers, with one characteristic dimension in nanometer scale. Material scientists have discovered unique properties of nanomaterials based on quantum size effects [344], while engineering scientists have developed specialized techniques for processing on the nanometer scale.Nano-scale synthesis ,nano-scale self assembly ,size evaluation and size reduction ,templating and lithography are examples of such techniques as summarized in Table 6.

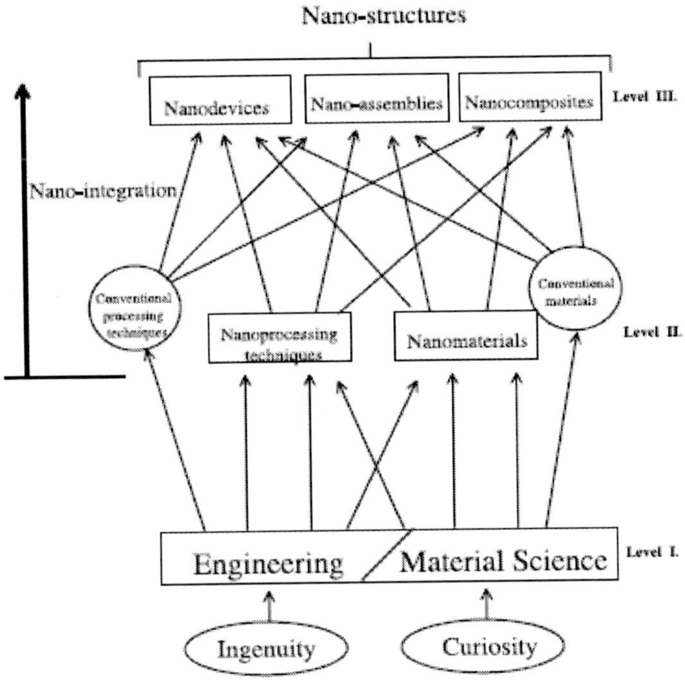

Typical nanomaterials.

- clusters
- nanoparticles
- fullerenes
- carbon nanotubes
- atomic and molecular layers on nanometer scale
- J-aggregates
- p-conjugated oligomers
- DNA molecules

Figure 76. outline of nano integration.

The goal of nano-integration is to construct nanostructures such as nanodevices, nano-assemblies and nanocomposites.In order to construct nanostructures nanomaterials as well as conventional materials must be used as the building blocks and nano processing techniques as well as conventional processing techniques, Table.6, must be utilized as building architecture. In other words, nano-integration is a process that starts from nanoscopic material and with the combination of conventional and nano processing techniques goes into the final of building nanostructures ,devices and assemblies.

Table 6. nano processing techniques as well as conventional processing techniques

• nanometer scale e-beam lithography	• ion-implantation
• optical near-field lithography	• chemical vapor deposition
• scanning probe techniques (AFM, STM)	• sputtering
• atomic layer epitaxyal growth	• etching
• Langmuir-Blodgett film growth	• film deposition
• cluster beam technique	• laser processing
• nanoparticle beam technique	• optical processing
• cluster size-selection technique	• molecular design and synthesis of polymers
• nanoparticle size-selection technique	• molecular design and modification of bio-polymers
• electric mobility utilization	
• molecular recognition	
• synthesis of mono-dispersed nanoparticles	
• structure formation by self-organization	
• nano-imprint lithography	
• templating technique	
• lithography-induced self-organization	

10.1. Nanoelectronics

The advantages of nanoparticle-catalyzed nanowire growth include its general applicability to a wide variety of materials; the diameter control affords the uniformity of the wires and their oriented, near-single crystallinity. The surface passivation, solubility and length the wires may also be systematically varied. Small-diameter "quantum" wires are ideal specimens for fundamental studies of two-dimensional (2D) quantum-confinement phenomena and for property comparisons to 3D-confined quantum dots, 1D-confined quantum wells and anisotropically 3D-confined quantum rods. Potential applications of semiconductor nanowires in nanophotonics and lasing, nanoelectronics, solar-energy conversion and chemical detection are under active development. Exciting advances are in progress in the semiconductor-nanowire field, enabled by the emergence of nanoparticle-catalyzed growth.The ability to create high-efficiency solar cells is a key strategy to meeting growing world energy needs. Nanotechnology is currently enabling the production of high-efficiency organic photovoltaics (OPVs) to help meet this challenge [345]. Polymer nanowires hold considerable promise for use in nanoelectronics and have already found application in practice. For example, thin films consisting of PANI nanofibrils synthesized by the interphase polymerization method are used as highly sensitive and fast detectors of vapors of acids (HCl) and alkalis (NH4OH). Conducting nanowires of polymers, such as poly(propylene) PP, PANI, and PEDOT (~100–200 nm in diameter), synthesized by the electrochemical polymerization method using Al2O3-based masks with nanopores show considerable promise for use as nanoemitters in emission displays. In a number of works, it has been shown that the electrical properties of polymer fibrils can be effectively studied using the field-effect transistor(FET) configuration, Figure 77.

In this configuration, the conductivity of polymer fibrils can be varied by varying the voltage applied to the gate. An analogous method has been used to study the carrier transporting doped PANI/PEO [345], andpoly-3-hexyl-thiophene (P3HT) [346] nanofibrils. It was shown in the above studies that the *I–V* characteristics of these transistor structures are nonlinear, as are the *I–V* curves of individual PANI fibrils (Figure 10.(ii) (b). These *I–V* characteristics differ from those observed in semiconducting [347] or organic [348-349] field-effect transistors in that the latter exhibit saturation of the current with increasing voltage. However, analogous nonlinear I–*V* characteristics have been observed in SWNT-based field-effect transistors [350], thin polymer strips ,SWNTs oriented using DNA [351], and a number of low-dimensional structures. The field mobility of carriers in a polymer fibril μFET can be estimated using the low-voltage segment of the *I–V* curve in the same manner as is done using the linear segment of the *I–V* curve of a traditional MOSFET Figure 77(c) . The carrier mobility μFET can be calculated from the transistor conductance *gm* [347]

$$g_m = \partial I_{ds}/\partial V_g|_{V_{ds}=\text{const}} = -(Z/L)\mu_{FET}C_iV_{ds}, \quad \textbf{(1.3)}$$

The fabrication of the electrode structure used for the growth of the conducting polymer wires is the same as for Pd fabrication. 2 μ liters of deoxygenated pyrrole (0.06 M in 0.01 M KCl) or aniline (0.1 M in 0.1 M HCl) solutions are used as the electrolytes . These electrolytes are placed in the e-beam patterned channel between the two electrodes. The electro deposition was carried out in galvanostatic mode by applying 100 nA current while the potential of the working electrode was monitored continuously with respect to a pseudo reference electrode. At the same time, a programmed multimeter is used to record resistances of the growing nanowires. he potential changed from the open circuit potential to a value of 2 V followed by a gradual decreased as the wire grew. When the wire was fully grown and made a contact with the cathode, the potential dropped to 0 V and the electro deposition process was stopped. Finally, the wafer was rinsed three times with deionised water. The measured resistances of grown conducting polymer nanowires were in the range of a few Kilo-ohms to Mega-ohms.

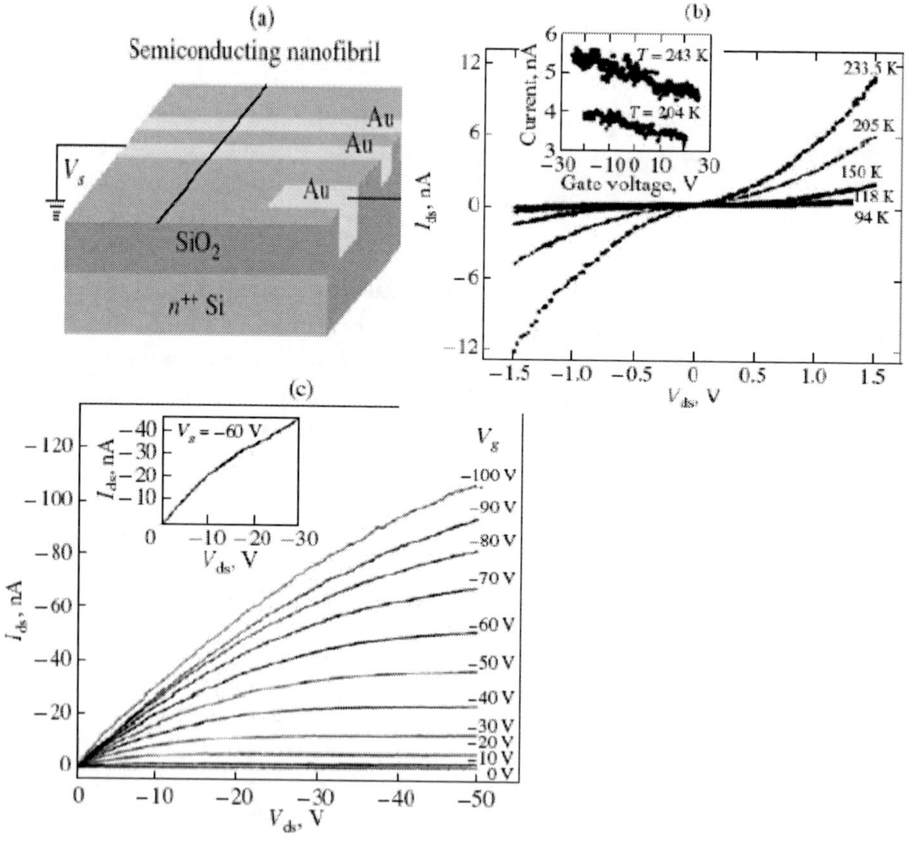

Figure 77. (a) Field effect transistor (FET) based on individual PANI fibril(b) IV characteristics of FET based on polyaniline fibril(20 nm) (c) IV characteristics of P3HT fibril.

10.1.1. Photovoltaics

Organic photovoltaics are nanostructured thin films composed of layers of semiconductor organic materials (polymers or oligomers) that absorb photons from the solar spectrum. These devices hold promise to revolutionize solar energy harvesting, because they can be manufactured via solution-based methods, such as ink-jet or screen printing, enabling rapid mass-production while scaling down cost. In recent years, bulk hetero-junction polymer photovoltaic devices and photodiodes have been constructed with composite films of CPs as electron donors and CNs as electron acceptors. The primary step in these polymer photovoltaic devices is an ultra-fast photo-induced electron transfer reaction at the donor-acceptor interface, which results in a meta-stable charge-separated state. In the case of the oligo (phenylenevinylene)/C_{60} composite, the quantum efficiency of this step is assumed to be close to one [352-353]. However, the overall conversion efficiency of these devices is limited by the carrier collection efficiency, which is greatly influenced by the morphology of the active film. Optical and photovoltaic properties of a composite based on MWNTs and PPV was studied by *Curran et al.*[354]. These authors claim that the helical structure of the polymer chain helps the MWNTs disperse in the polymer solution, and that LEDs made with this composite are five times more stable in air compared with devices without the MWNTs. However, when one utilizes the composites for electronic devices, the nanotubes can cause a short-circuit due to their high conductivity. *Romero et al* [355] fabricated diodes with double layers of PPV and MWNTs, but the MWNT film, which was prepared by filtration, was so rough that a very thick PPV layer was needed to avoid a short-circuit. The composite was realized by spin-coating the PPV precursor on thin MWNT films and subsequent conversion at high temperature. A drastic reduction of the PL efficiency and a change in the vibration structure of the PL spectrum were affected by MWNTs.The reduction of the PL efficiency was understood as resulting from an inter-system energy transfer (singlet-triplet), a partial hole transfer from PPV chains to MWNTs and both superposing on an intense scattering and absorption of the exciting light by MWNTs.Using this composite, photovoltaic devices have been fabricated by employing MWNT as a hole-collecting electrode. A good quantum efficiency (1.8% at 2.9–3.2 eV), about twice that of the standard ITO device was obtained. The high efficiency was considered arising from a complex interpenetrating network of PPV chains with MWNTs and the relatively high work function of the MWNT film. The reported results have suggested the possible application of CNs as a new interesting electrode material in macro scale devices. SWNT/poly(3-octylthiophene) (P3OT) composites have been used for the fabrication of new photovoltaic devices

[356].P3OT, acting as the photo excited electron donor, is blended with SWNTs which act as the electron acceptors. In such devices the transferred electrons are transported by percolation paths provided by the addition of SWNTs.Diodes (Al/polymer-nanotube composite/ITO) with a low nanotube concentration (<1%) have shown a photovoltaic behavior with an open circuit voltage of 0.7–0.9 V. The short circuit current was increased by two orders of magnitude compared with the pristine polymer diodes and the fill factor also increased from 0.3 to 0.4 for the nanotube/polymer cells. With the above observations in view, the authors proposed a photo-induced electron transfer at the polymer/nanotube interface was proposed . It was shown that the internal polymer/nanotube junctions act as dissociation centers, which were able to split up the excitons and also created a continuous pathway for the electrons to be efficiently transported to the negative electrode. This resulted in an increase of electron mobility, and hence, balanced the charge carrier transport to the electrodes; the conductivity of the composite was increased by a factor of 10, indicating percolation paths within the materials. A conclusion of these results was that the CP/SWNTs composite represented an alternative class of hybrid organic semiconducting material that hold promise for organic photovoltaic cells with improved performance. Nanocomposite materials are of interest in recent years for their potential applications in electronic devices such as organic light emitting diodes (OLEDs) and photovoltaic cell. In the simplest version an OLED consists of a layer of an electroluminescent organic material sandwiched between two electrodes. One of the electrodes must be transparent to transmit light created during the electroluminescent effect. The luminescent emission of OLEDs is due to the radiative recombination of excitons.Fabrication of high efficient OLEDs depends not only on the electronic and the optical properties of the pure organic materials but also on the control of charge transport, holes or electrons through the buffer layers used and the enhancement of charges migration by doping the emissive material [357-358].A proper layer combination in OLEDs can also balance the injected charges in an emissive layer thus increasing the external efficiency. The use of a buffer layer leads to a reduction of the charge injection barrier and an even charge distribution with a large contact area at the interface between the electrodes and organic materials. A recent work shows that the dispersion of SWNTs in a host polymer (*m*-phenylenevinylene-*co*-2,5-dioctoxy-*p*-phenlenevinylene) (PmPV) (hole conducting) traps the holes, depending on the applied voltage, in a double emitting organic light emitting diode (DE-OLED)[359].This hole trapping leads to the changing of the radiative recombination regions in the DE-OLED. The tris-(8-hydrozyquinolinolato) aluminium (Alq3) doped by Nile Red was used as an emissive material between the polymer composite and cathode. The device

fabricated without SWNTs dispersed in the PmPV has shown a dominant emission near red at 600 nm, which is in the range of the characteristic emission of Nile Red-doped Alq3, while the addition of a small amount of SWNTs enhanced a green emission. The devices fabricated with the polymer/SWNTs composite have shown an increase in the oscillator strength of the green emission with a dominant emission peak near 500 nm, the characteristic emission of PmPV.This fact was observed for SWNT concentrations up to 0.1 wt%. The shift in the emission indicated that the SWNTs in the PmPV matrix act as a hole-blocking material that results in a shifting of the recombination region from the Nile Red-doped Alq^3 layer to the PmPV composite layer .*Romero et al.*[355] studied the hetero-junction of CNs with an organic semiconductor. This polymer/CN composite was expected to create a polymer-nanostructure matrix with cooperative behavior between the host and additive thus modifying the electronic properties of the polymer composite. It is worth noticing that in the CP field, PEDOT is known as one of the best hole conducting buffers since its ionization potential is close to the work function of ITO while its electron affinity, of near 2.2 eV, is small enough to block electrons. For this reason, it has been interesting to investigate the electronic properties of SWNTs dispersed in PEDOT since the pure SWNTs have experimentally been shown to be a *p*-type semiconductor. In order to investigate the role of SWNTs in a hole CP, OLEDs were fabricated with a conjugated emissive copolymer, poly (3,6-N-2-ethylhexyl carbazolyl cyanoterephtalidence) (PECCP) and SWNTs dispersed in a hole conducting buffer polymer, PEDOT [360]. Devices made with SWNTs dispersed in PEDOT and devices made without addition of SWNTs in the PEDOT emit green light at 2.37 eV as expected for PECCP. It was observed that the device made with SWNTs in the buffer layer showed a significant decrease in the electroluminescence (EL) as compared to that of the device without the SWNTs.The photoluminescence (PL) from the same organic layer combination, excited from the PECCP side and measured through the PEDOT and the ITO glass, showed a very little difference between the films with and without the SWNTs.The current-voltage (I-V) characteristic of OLEDs with SWNTs has illustrated a lower I-V power dependence (I-V) near 1–2 V than the device without SWNTs, which has a power dependence of I-V.The EL and the I-V data together with the PL have suggested an electronic interaction between the SWNTs and the host polymeric material, PEDOT.It has been proposed that this electronic interaction originates from the hole trapping nature of SWNTs in a hole CP. *Kim et al.* [361] devoted a paper concerning the fabrication of OLEDs with the structure of ITO-coated glass/PEDOT:PSS/SWNTs-PVK (poly-carbazole) nanocomposite /(4-(dicyano methylene)-2-methyl-6-(p-dimethyl aminostryryl)-

4H-pyran (DCM)-doped) Alq3-Li Al.SWNTs were dispersed with the hole CP and PVK in toluene .SWNTs-PVK nanocomposite was spin-coated on the PEDOT:PSS-ITO-coated glass and then DCM-doped Alq3 was deposited by cluster beam deposition methods. The device qualities such as external quantum efficiency were improved by a factor of 2–3 for SWNTs concentration up to 0.2 wt%.

10.1.2. Solar Cells

Solar cells are devices, which operate inversely to OLEDs, they transform radiation from the optical range into electricity. Photon absorption in the organic-based composites produces primarily bound-state excitons.Dissociation of these charge pairs is facilitated by the potential difference across a polymer-metal junction, provided by agglomeration of excitons near the interface. The dissociation can be accomplished via electron acceptor impurities, too [362]. Under illumination, a transfer of electrons to the acceptors will take place and the holes will be preferentially transported through the CP.This process is known as photo-induced charge transfer. Since the discovery of photo-induced charge transfer, a variety of acceptor materials have been introduced into CPs to produce photovoltaic devices (for example fullerene and SWNTs) [363-364].The use of SWNTs in this device is a particularly attractive approach for several reasons. At low doping levels, percolation pathways are established, providing the means for high carrier mobility and efficient charge transfer. This has been a problem in the majority of polymer solar cells developed to date, even with the advent of semiconductor nano-rods. Since the diffusion distances for excitons in CPs, for example PPV, have been reported at <10 nm, the requirement for a sufficient percolation network of electron-accepting dopants in the polymer composite is substantiated [364].Electrical conductivity data has validated that SWNTs-doped polymer composites demonstrate this extremely low percolation threshold. Other beneficial properties of SWNTs relevant to polymeric photovoltaic development include composite reinforcement and thermal management.SWNTs have shown promises in the development of CP/NC composites with enhanced mechanical strength by load transfer from the polymer matrix to the dopant[365]. Tensile strengths of SWNTs have been estimated to be ca.20 GPa and the Young's modulus measured by atomic force microscopy was of ca. 1 TPa [366].This high Young's modulus and strength/weight ratio could help provide much-needed mechanical stability to large-area thin-film arrays. SWNTs may provide assistance in thermal management for such arrays, too.The thermal conductivity of an isolated SWNTs has been theoretically predicted to be as high as 6600 W/mK

[367].Polymer composites doped with as little as 1% wt SWNTs have shown a 70% increase in the thermal conductivity at 40 K [368].Future optimization of carrier transport and exciton dissociation with this novel photovoltaic system has been proposed to derive from incorporation of nanostructure-SWNT complexes into the polymer.

10.1.2.1. Dye-Sensitized Cells

These cells employ complex dye molecules attached to the surface of nanostructure oxides like titanium(IV) oxide (TiO2) or niobium(V) oxide (Nb2O5). The dyes exhibit broad light absorption profiles and rapid photo induced charge transfer of electrons to the nanocrystals. These cells show solar conversion efficiencies of ~4%.

10.1.3. Sensors

Recently, CN-based gas sensors have received a great deal of attention.Nanosized CN-based gas sensors of the field effect- transistor have superb sensitivity at room temperature due to a drastic change in the electrical conductivity upon the adsorption of various gases [369-370].Despite such advantages, however, their application is still limited by a long recovery time and a complex fabrication process. According to *Qi et al.* [371] SWNTs used for gas sensors must be semiconducting. The presence of both metallic and semiconducting CNs in conventional powder samples reduces the reproducibility and/or yield of the devices. The first application of CN/CP nanocomposites was reported in 2004, when a gas sensor was fabricated based on a SWNTs/PPY composite using a simple chemical polymerization followed by spin-casting onto pre-patterned electrodes [372]. This method is a simple way to form a uniform coating of PPy on SWNTs.It was investigated as being obtained an SWNT sensor with *n*-type behavior. The authors also showed that the prepared PPy and SWNT/PPy composite had an *n*-type behavior, which was explained by anion doping in PPy during the chemical polymerization process. The sensitivity of the gas sensor fabricated with the SWNT/PPy nanocomposite towards NO_2, as measured by a direct voltage divider at room temperature, was very high of about ten times higher than that of PPy done nano-dispersing of the SWNT bundles, an increase of the specific surface area of the coated PPy took place and thereby a further increase of the sensitivity. *Valentini et al.* [373] reported in the same year a higher sensitivity of a sensor based on poly (o-anisidine) (POAS) deposited onto CNs.Upon exposure to inorganic vapour (for example HCl) the variation of the CNs sensitivity is less than 4%, while the POAS-coated CNs devices offer a higher sensitivity (i.e., 28%). The extended detection capability to inorganic vapors is attributed to direct charge transfer with electron hopping effects on inter tube conductivity through physically adsorbed POAS between CNs.In the meantime, using chemical polymerization *Bavastrello et al.* [373-374] reported a new nanocomposite based on poly (2,5- dimethylaniline) and MWNTs aiming at its application in conductometric acid vapors sensors. These authors put in evidence a progressive spontaneous undoping process with time associated with the instability of the doping agent.. In their opinion, this phenomenon is related to the synergetic effect of the steric hindrance of the substituents on the aromatic rings and the presence of MWNTs inside the polymer matrix [374] also described the fabrication of a spontaneous reversible sensor for acid vapors by setting up a comparative potentiometric circuit and engineering the sensitive element directly onto the circuit board.

10.1.4. Biosensors

Electrochemical biosensors are modified electrode devices that rely on the intimate coupling of a biomolecule (most often an enzyme) and an electrode transducer that converts a specific biological recognition event into a measurable electrical signal. Successful operation of biosensors involves therefore effective immobilization of the biomolecule while retaining its activity over prolonged periods of time, as well as its affinity for and accessibility towards the target analyte. Many amperometric biosensors, particularly the oxidase-based ones, require the use of charge-transfer mediators or electro catalysts. To be efficient, the latter should be immobilized in the close vicinity of the electrode surface while keeping mobile enough to facilitate communication between the active sites of the biomolecule and the electrode surface. In general, carbon materials are considered good transducers, stabilizer and mediators for biosensor applications [375].As far as we know, so far CNs based materials have been used to improve more than ten times the transducing magnitude of catalytic biosensors, to immobilize the proteins thanks to the strong interaction between the enzyme and the nanotube surface to promote redox reaction and to enhance the electron transfer in HEME proteins, even overcoming the lack of specific binding of CNs to proteins using polymer chains[376]. The use of MWNTs provides a novel electrode platform for CPs-based biosensors [377].An example in this sense is the PPy based glucose oxidase system for detection of glucose. The use of these three-dimensional electrodes offers advantages in that large accessible enzyme loadings can be obtained within an ultra-thin layer. Another example in the biosensors field is that of the DNA-doped PPy film coated on the CN modified electrode which was used to the free DNA hybridization detection by impedance measurements [378].Simple indicator or label free DNA hybridization strategy represents a very attractive approach but has been difficult to achieve, particularly for electrochemical devices, since DNA duplex formation does not directly lead to a change in redox signal.CPs, such as PPy, allow for the intimate association between a biological recognition element and a potential reporter polymer chain. A new biosensing strategy for direct electrochemical detection of DNA hybridization by AC impedance measurement was described [378]. The indicator or label free approach was developed on the basis of the CN coupled with PPy. MWNTs functionalized with carboxylic group (MWNTs-COOH) were modified on the glassy carbon electrode (GCE) and the oligo-nucleotide probes were doped with the electro polymerized PPy films by serving as the sole counter anion during the growth of the conducting films. Before and after hybridization reaction with the complementary DNA sequences, a decrease of impedance values was observed as a consequence of the reduction of the electrode resistance.Such unique response was attributed to the concomitant conductivity changes of the PPy-polymerized CNs, and offers great promise for reagent less DNA hybridization analysis. The method has many advantages, such as reducing reaction time without using any indicator or fluorescent materials and high selectivity for complementary and mismatched target sequences. A recent impedance study has shown that the addition of MWNTs to a CP matrix results in the best decrease of the charge-transfer resistance and of the mass transfer impedance [379].This fact demonstrates that materials based on CNs are the best improved polymeric nanocomposites for biosensor applications because they provide the best electron transfer and assure a faster ion mass transfer.

10.1.5. Hydrogen Storage

Cho et al.[380] reported approximately 8 wt% (reversible) hydrogen gas storage in doped forms of the organic conducting polymers such as polyaniline and polypyrrole. There has been, and still is, some confusion among investigators who are not familiar with polyaniline. Polyaniline does not exist as such. It exist in 3-different discrete oxidation states: (i) the fully oxidized pernigraniline oxidation state, (ii) the half oxidized emeraldine oxidation state and (iii) the completely reduced leuco oxidation state, or in oxidation states intermediate between any two of the above adjacent oxidation states.

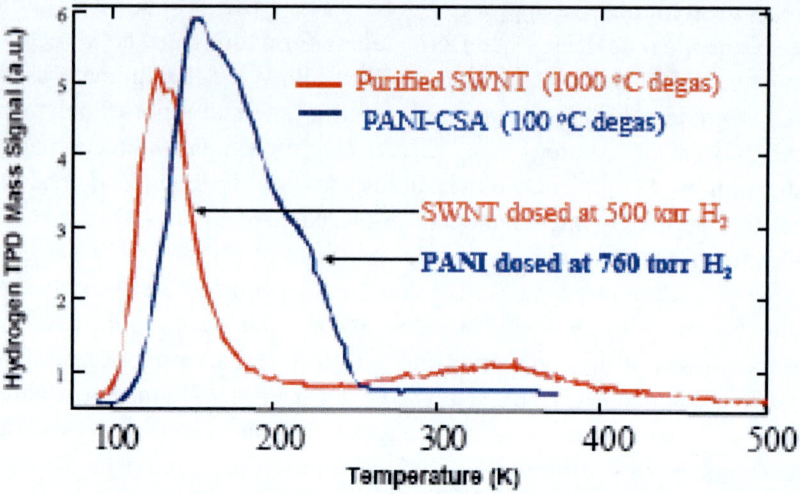

Figure 78. Desorption studies of PANI nanofibers doped with CSA.

The unique nature of this fascinating class of compounds highlights its potential importance in a new type of H_2 reaction, e.g. adsorption of H_2 , especially since it can also be prepared in many different nano/micro morphology [2,3]. There is experimental evidence for an interaction between H_2 and doped polyaniline. This interaction was certainly an electronic interaction as shown in the change of electrical resistance measurements of PANI/ CSA in the presence of hydrogen (H_2), Figure 78.

10.2. Nanomedicine

Nanomedicine has been defined as ''the monitoring, repair, construction, and control of human biological systems at the molecular level, using engineered nanodevices and nanostructures''[381]. Therefore, nanomedicine adopts the concepts of nanoscale manipulation and assembly to applications at the clinical level of medical sciences. In a broad sense, nanomedicine is the application of nanoscale technologies to the practice of medicine. It is used for the diagnosis, prevention, and treatment of disease and to gain an increased understanding of complex underlying disease mechanisms[382]. Although nanotechnology is an established discipline, commercial nanomedicine (with its broad range of ideas, hypotheses, concepts, and undeveloped clinical devices) is still at a nascent stage of development. For example, there are many nanodevices (eg, Q-dots, dendrimers) that are widespread and broadly marketed, but have yet to find their way into a wide range of clinical devices. This is a consequence of the extremely complex and demanding requirements of clinical trials by the FDA, which can take years before a product makes the long trek from a concept in the laboratory to a commercially viable medical product for the consumer[383].Currently, nanomedicine involves detection of particles , drug delivery systems, emulsions, and carriers for delivering vaccines; and nanofabricated biomaterials with unusual properties of strength, hardness, reduced friction, and improved biocompatibility. More exotic concepts (eg, nanomachines that could move through the body; troubleshooting and repairing tiny brain or cardiovascular lesions) lie in the future.

10.2.1. Biomedical Applications

The appearance of conducting polymers in the biological field only happened in the 1990s. One major area is the development of biosensors based on the fact that CPs can be switched from conducting to insulating or back upon contact with certain compounds (e.g. an electronic nose to detect odor)[384].Another fact being exploited is that special biomolecules, like growth factor or enzyme, could be incorporated as dopant into the structure of CPs. As a result, the CP can either promote specific cell growth or catalyze a bioreaction. Then by tracking the catalyzed reaction, one could detect the concentration of certain molecules (e.g. glucose, cholesterol, penicillin)[385]. However, the low sensitivity of CPs based sensors and conductivity susceptible to humidity limited the commercialization. *In vitro* and *in vivo* studies, however, brought the biocompatibility issue of CPs.There is not a large amount of data in this area, though some *in vitro* experiments have been carried out without profound analysis of the biocompatibility first[386-387].Tissue biocompatibility has been proved for polyaniline films in all emeraldine (EM), nigraniline (NA), and leucoemeraldine (LM) forms through subcutaneous implantation into rats[388]. On the other hand, *in vitro* cellular response to polypyrrole coated woven fabrics (Context TM),according to Barbara Jakubiec, was not so good[389].High levels of conductivity are detrimental to the growth, migration, and viability of endothelial cells while lower conductivity (103 ~104Ω/ square) is better in terms of cellular responses. Despite the differences in the type of CPs used in the two studies, it is worth mentioning that cells are generally more sensitive to the surroundings thus *in vivo* tissue biocompatibility cannot guarantee *in vitro* cyto-biocompatibility. In fact, an organism (animal or human body) can trigger a series of host reactions with the body's defense mechanism towards the foreign material to survive the attack. *In vitro* applications of conducting polymers are, so far, around electrical stimulation of cell growth. With nerve cells, results due to the electrical stimulation result in the increased protein adsorption (fibronectin) and enhanced neurite length [387,390].With endothelial cells, it has been found, even without stimulation, cell extension as well as DNA synthesis could be non-invasively controlled by switching the CP from oxidized state to neutral state. Yet conclusive data concerning bone cells stimulated by CP constructed electrical field or current are rare. A conductive nanocomposite made of Poly-Lactic acid and carbon nanotubes was used as a substrate to stimulate osteoblasts growth [391]. Increased proliferation, concentration of calcium in extra cellular matrix, and up regulation of mRNA expression for collage type I were found when cells were exposed to 10Hz electric stimulation. The conductive material used in this study is not an intrinsically conducting polymer, though.

10.2.2. Drug Delivery

The ability to create materials with well-controlled structures on the nanometer length scale is of intense interest for a variety of applications,[392-393] including controlled drug delivery[394] and biomedical devices.[395-396]. Conducting polymers are of considerable interest for a variety of biomedical applications.[396] Their response to electrochemical oxidation or reduction can produce a change in conductivity, color and volume [397-398].A change in the electronic charge is accompanied by an equivalent change in the ionic charge, which requires mass transport between the polymer and electrolyte.[399]. When counter ions enter a polymer it expands and when they exit it contracts. The extent of expansion or contraction depends on the number and size of ions exchanged [400]. Electrochemical actuators using conducting polymers based on this principle have been developed by several investigators.[401-403]. They can be doped with bioactive drugs, and can be used in actuators such as micro fluidic pumps.[404-405] .The precisely controlled local release of anti-inflammatory drugs at desired points in time is important for treating the inflammatory response of neural prosthetic devices in the central and peripheral nervous systems.[406]. Here we report on a method to prepare conducting-polymer nanotubes that can be used for precisely controlled drug release.

The conducting-polymer nanotubes significantly decrease the impedance and increase the charge capacity of the recording electrode sites on micro fabricated neural prosthetic devices.The drugs can be released from the nanotubes in a desired fashion by electrical stimulation of the nanotubes; this process presumably proceeds by a local dilation of the tube that then promotes mass transport. Microelectrode neural probes facilitate the functional stimulation or recording of neurons in the central nervous system and peripheral nervous system. Conducting polymers such as polypyrrole (PPy) and poly(3,4-ethylenedioxythiophene) (PEDOT) were in templated nanostructures for neural prosthesis applications.[407-408]

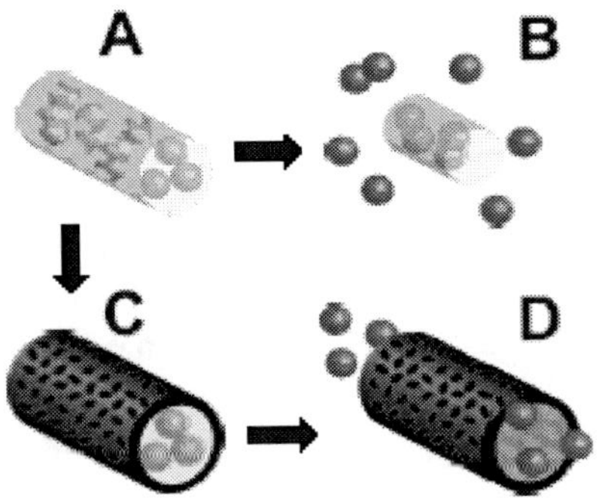

Figure 79. Schematic illustration of the controlled release of dexamethasone.

Minimizing the electrode impedance is an important requirement for obtaining high quality signals (high signal-to-noise ratio).[409] PPy and PEDOT decreased the impedance of the recording sites.[407-408]. PEDOT nanotubes that have a well-defined internal and external surface texture further decrease the electrode impedance by increasing the effective surface area for ionic-to-electronic charge transfer to occur at the interface between brain tissue and the recording site. The release of dexamethasone, could be precisely controlled by external electrical stimulation of PEDOT nanotubes. In order to produce the nano tubular conducting polymers, nanofibers of biodegradable poly(L-lactide) (PLLA) or poly(lactide-*co*-glycolide) (PLGA) were first electrospun onto the surface of a neural probe followed by electrochemical deposition of conducting polymers around the electrospun nanofibers. In a final step, the fiber templates could be removed or allowed to slowly degrade, providing additional means of controlled delivery of biologically active agents incorporated into the fibers themselves , Figure 79, PLLA and PLGA were considered as suitable polymers for the template since they can be readily processed into nanoscale fibers, are stable during electrochemical deposition of the conducting-polymer coating, and can be easily removed under conditions that leave the wall material intact.[410] PLLA/ PLGA nanoscale fibers and dexamethasone-incorporated PLGA nanoscale fibers were collected on the micro fabricated

10.2.3. Conducting-Polymer Nanowire Immunosensor

A promising approach for the direct (label-free) electrical detection of biological macromolecules uses one-dimensional (1-D) nanostructures such as nanowires and nanotubes, configure d as field-effect transistors that change conductance upon binding of charged macromolecules to receptors linked to the device surfaces. Combined with simple, rapid, and label-free detection, potentially to single molecule, these nanosensors are also attractive due to the small size, low power requirement and, most of all, the possibility of developing high-density arrays for simultaneous analyses of multiple species. Although current nanosensors based on carbon nanotubes and silicon nanowires have elucidated the power of 1-D nanostructures as biosensors, they have low throughput and limited controllability and are unattractive for fabrication of high-density sensor arrays. More importantly, surface modifications, typically required to incorporate specific antibodies, have to be performed post synthesis and post-assembly, limiting the ability to individually address each nanostructured sensing elements with the desired specificity,Figure 80.

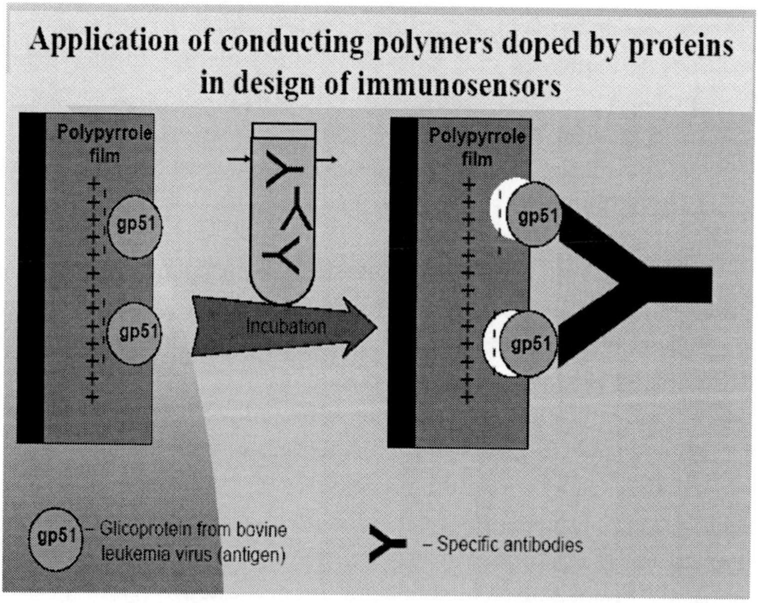

Figure 80. Application of conducting polymers in the design of immunosensors.

A novel technique has been developed for the facile fabrication of bioreceptor (antibody)-functionalized nanowires that are individually addressable and scalable to high-density biosensor arrays and to demonstrate its application for label-free, real-time, rapid, sensitive and cost-effective detection of multiple pathogens in water. Electropolymerization of conducting polymers between two contact electrodes is a versatile method for fabricating nanowire biosensor arrays with the required controllability. The benign conditions of electropolymerization enabled the sequential deposition of conducting-polymer nanowires with embedded antibodies onto a patterned electrode platform, providing a revolutionary route to create a "truly" high-density and individually addressable nanowire biosensor arrays. The nanowire immunosensor array's utility could be used to simultaneously quantify three important model pathogens: poliovirus, hepatitis-A virus and rotarvirus. A facile technique of electrochemical polymerization of biomolecule-friendly conducting polymers, such as polypyrrole, in prefabricated channels of tailor-made aspect ratio between two contact electrodes at site-specific position to synthesize nanowires of tailor-made properties for fabricating individually addressable high-density nanowires biosensors arrays,Figure 80 [411-413]. Detection of pathogens was achieved by the extremely sensitive modulation of the electrical conductance of the nanowires brought about by the change in the electrostatic charges from binding of the pathogens to the antibodies. Effects of monomer concentration, dopant type and concentration, aspect ratio and electrochemical polymerization mode on the sensitivity, selectivity and durability of poliovirus, HAV and rotavirus antibodies functionalized polypyrrole nanowires as label-free bio-affinity sensor of these important model viral pathogens in water were investigated to establish optimum synthesis conditions of biomolecules-functionalized nanowires [413].

10.2.4. Enzyme Encapsulation

Nanoscale wires are of interest as components of "electronic nose" sensor and biosensor arrays. Arrays of macroscopic sensors can be used for such diverse applications as determining breath alcohol levels, automotive exhaust safety inspection, and identification of toxic gases[414].The primary advantage of nanosensors over their macroscopic counterparts is their ability to detect analytes in a power- and area-efficient manner, and to be integrated conveniently into chips that combine the functions of chemical sensing and signal processing. Because of their high specificity for both small molecules and biological analytes, as well as their ability to amplify signals through catalytic reactions, enzymes are key components of these nanosensors[415-416].A template-based approach was developed to encapsulate active enzymes in conducting polymer "stripes" within metallic nanowires,Figure 81. These nanowires were integrated into test circuits and lithographically patterned sensor arrays suspending them in water and applying an AC voltage to the contact surfaces. Proteins (avidin, streptavidin, catalase) in the polymer segments retained their molecular recognition and catalytic functions.

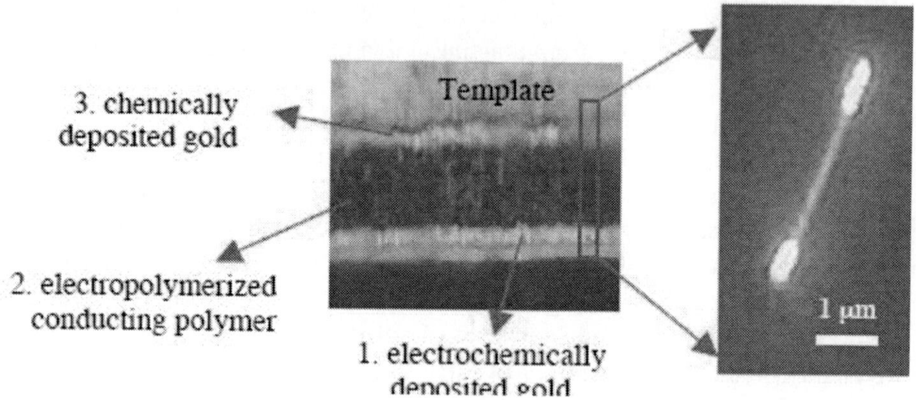

Figure 81. electrochemical replication of porous templates (in this case an anodic alumina membrane) free-standing nanowires containing conducting polymer embedded proteins between metal end contacts.

The conjugation of biotin to polymer-bound avidin provided a "hook" for binding other specific sensory or catalytic receptors[417]. In addition to their sensory function, these segmented nanowires are interesting as objects that move autonomously in "fuel" solutions by catalyzing reactions at one end. The incorporation of catalytically active enzymes into polymer segments opens the interesting possibility of using bio-friendly fuels such as glucose to power these nanoscale motors[417].

10.2.5. Antibacterial Agents in Packaging Systems

Smart packaging' refers to the use of barrier materials as packaging film to improve shelf life which include , moisture absorbent sachets, oxygen scavengers, transport shock resistant materials , time and temperature detectors i.e. for thawed frozen goods. Nanostructured antimicrobial coatings fall under this category. Conducting polymers (artificial metals) is a big area of new research because these materials potentially allow the printing of transistors and semi-conducting devices right onto the pack. Combined with a printed power supply and presto you have a pack that can literally communicate with the consumer. Thermochromic, electrochromic and chemo chromic inks based on conducting polymers are able to sense and communicate, and other printing techniques provide optical patterns (holograms and other optical variable devices

(OVDs).PANI based on chemical detector can detect changes in meat and dairy products.The PANI reacts to pH changes and oxidation and changes color. It commonly exists in protonated (green) and unprotonated (blue) forms. In tests the sensors were used to detect the spoilage of milk. Polyaniline in its blue form was grafted on acetate and left for several days. As the milk went sour, lactic acid formed and this interacted with the polyaniline which changed from blue to green. The sensor was also tested with packaged meat. Polyaniline in its green form was grafted onto acetate and placed in a sealed container with a sample of ham and left for several days. Various amines such as ammonia formed as the meat began to degrade. The amines interacted with the polyaniline and there was a clear color change from green to blue. Its commercial viability is still under question, but it does demonstrate that food safe polymers are available to give indications of chemical change. Chemical changes can also result in off-gas other than oxygen or carbon dioxide, which has detectable changes to certain sensors.

10.2.6. Actuators

Actuators directly converse electrical energy to mechanical energy with a material response to applied electrical voltage. Due to these characteristics that are similar to biological muscles, actuators have been dubbed as "artificial muscles". Applications of actuators have been found in robotics, optical fibre switches, optical displays, prosthetic devices, microscopic pumps and anti-vibration systems. Electro active conducting polymer (CP) actuators operating on electrochemical dopant intercalation were proposed a decade ago. One of the attractive features of the CP actuators is low drive voltage that does not exceed 5 V that is much lower value compared with conventional materials such as ferroelectric and electrorestrictive materials. Even at low voltage, in suitable conditions CP actuators can reach an actuation strain as high as 7% generating stresses that are comparable to or exceed those of biological muscles. Recently, carbon nanotubes (CNT) have been proposed as actuator material.CNT actuators have shown actuation strains in the order of 0.2 % with drive voltage of few volts.CNT and electrically conducting polyaniline (PANI) nanocomposite as an actuator material have been developed. The actuation synergism of PANI and CNT as actuator materials and the ability of actuation in aqueous salt solutions and acid solutions has been examined. Polypyrrole is one of the conducting which has been used extensively used as actuators muscles". In step -1 a layered structure of Si/SiO2/Ti/Al is fabricated. In step 2 anodic etching is used to oxide aluminium to alumina porous membrane capped by a barrier layer of Ti.This barrier layered is then removed by reactive plasma etching and PPy is electrochemically deposited into the pores. The alumina membrane is then chemically removed leaving free standing polypyrrole nanorods,Figure 82.

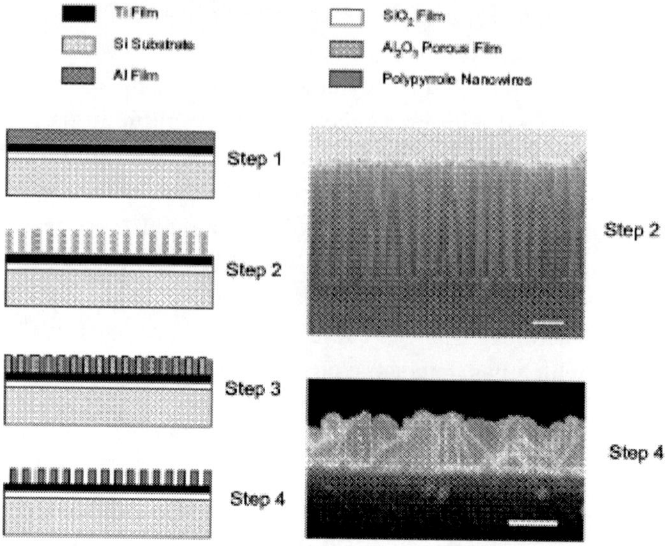

Figure 82. Fabrication of polypyrrole nanorods.

11. Challenges in Processing and Manufacturing of Nanocomposites

In spite of the remarkable developments achieved in the preparation techniques of nanocomposites, there are still some problems associated with the liquid processes involved in nanocomposite preparation, such as the requirement of a large amount of aqueous/organic solvent, the aggregation of clay due to insufficient affinity with the monomer, and the lower level of dispersibility in matrix polymers.[418].But developing the processing–manufacturing technologies in terms of quantity and value for commercialization will be one of the biggest challenges. For example, dispersion of nanoparticles or chemical compatibility with matrix materials is the important issue. By using traditional techniques, a homogeneous dispersion of nanoparticles in a polymer is very difficult due to the strong tendency of fine particles to agglomerate [419-420]. At the same time if it is subjected to force, there is a possibility of splitting of agglomerate nanoparticle. Therefore, premature failure takes place in the final product [420]. The alignment of nanoparticles in the composite matrix can be critical to maximize unidirectional properties such as strength, modulus, and toughness [421-424].As in the case for traditional composites; it is even more challenging to determine the strength, composition and functionality of the interfacial region. The goal of improving the carbon nanofiber matrix interfacial adhesion issue and complete dispersion must be solved before achieving the full potential of nanocomposites. The technique holds promise for low viscosity resins if the temperature and/or added diluents help to reduce the viscosity of the mixture. In addition to the composite integrity, the nature of the nanoparticle is also critical for property improvement. For the generation of a model configurations of nanotubes as well as to improve dispersion and compatibility in polymer matrices, nanotube is being functionalized. There are still some concerns like whether the functionalization of a nanotube will affect the properties to improve the final product. In heterogeneous systems, size effects are influenced also by the chemical nature of components, morphology of particles, their spatial distribution, and interface interaction. Because of this, methods for manufacturing nano-sized heterogeneous systems can play important role in the properties of the resulting nano products. Inspite of their preponderance , manufacturing nanocomposites according to these technologies is rather complicated.

12. Toxicology/Safety of Nanoparticles and Nanostructures

There is a general concern that nanoscale particles may have negative health and environmental impacts [425]. Few studies have been published on the effects of inhaling free manufactured nanoparticles to date. Although nanometals are actually introduced into the body for disease detection, imaging, and treatment, there are concerns about the potential toxicity of such materials [426]. The exceptional properties of carbon nanoparticles may also lead to unique health hazards [427]. This is because the size of the nanomaterials is comparable to human cells and large proteins with the result that the regular human immune system may not work against them [428]. While nanoparticles may enter the body by inhalation, due to skin contact, or by ingestion, the major concern is by inhalation during processing and use and reuse of these ultrafine particles [429]. Humans continually come into contact with small particles, and they are generally not hurt by being exposed to these particles. However, inhalation of micron-sized quartz particles or of asbestos fibers has proven to be quite harmful to health [430]. Similarly, nanoparticles resulting from combustion processes such as forest fires or industrial pollution have a detrimental effect [430]. In addition, studies have been conducted to observe the pulmonary toxicity of nanosized particles in rats [431]. It has been found that toxicity is related to inhalation of nanoparticles, which can penetrate into deep lung tissue. It has been suggested that the surface properties such as surface chemistry and the area of nanoparticles and free radical generation with cell interaction play a significant role in the increased toxicity of nanoparticles [432]. Thus, the small size of the particles and the coating on these particles are both important considerations [429]. The adverse effects of carbon nanotubes (CNT) exposed to human epidermal keratinocytes (HaCaT) were evaluated [431]. *Moteiro Riveria* [433] has recently shown that MWNTs can localize within and initiate an irritation response in a target epithelial cell (human epidermal keratinocytes).Shvedova et al. [431] observed exposure of HaCaT cells to CNT resulted in accelerated oxidative stress and a reduction of total antioxidant reserve and vitamin E. In addition they showed ultra structural and morphological changes in cultured skin cells followed by carbon nanotube exposure. Deposition of ultrafine titanium dioxide (TiO2) particles from the sunscreens was observed in the human stratum corneum within hair follicle areas and deeper parts of the hair bulb and papilla [434]. There is some evidence indicating that nanoparticles of TiO_2 (used in some sun protection products) do not penetrate the skin but it is not clear whether the same conclusion holds for individuals whose skin have been damaged by sun or by common diseases such as eczema [422]. TiO_2 crystals (20–100 nm) were shown to be solely deposited on the outermost surface of the stratum corneum and were not detected in deeper stratum corneum layers of the human epidermis and dermis [431]. Chemically modified TiO2 nanoparticles from over-the-counter sunscreen products have been shown to induce the formation of hydroxyl radicals and oxidative DNA damage. Additionally, sunlight-illuminated TiO_2 was shown to cause catalyzed DNA damage in human cells [435]. Combustible nanoparticles may cause an increased risk of exposure as a function of their increased surface area [435]. There has been little discussion among researchers working on plastics about the toxicological aspects of these nanofillers even though polymer nanocomposites are being used in the auto industry [435]. Polymer–clay nanocomposites have been shown to possess flame-retardant properties. However, one needs to be aware of the possibility of nanosized particles being released into the atmosphere as a result of combustion and then being inhaled by humans [429]. For PNCs used in automobiles, this may happen during an accident when the vehicle may catch fire, and it may also happen at the end of the automobile's useful life, when the discarded PNC may end up in an incinerator as a waste fuel [429]. *Agarwal et al.* [429] studied PNCs containing different types of surface treated nanoclays and carbon nanofibers, and burned in a Combustion DMS 500 Differential Mobility Spectrometer. The results showed an increase in the rate of particle emission with a decrease in particle size distribution. Though in terms of safety it was not very well understood, these features are not desirable! Introduction of nano materials into industry requires safety evaluation as well as an understanding of the impact on the environment, and human health or other biological species [436]. This safety information has been worked out by industry and currently has not been widely published. *Fiorito et al.* [437] investigated highly purified carbon nanomaterials (SWNTs-C-nanotubes, C-fullerenes) and concluded that they possess a very low toxicity against human macrophage cells. As the determination of toxicity of C-nanoparticles is quite complicated [437], more work is needed on all of the new carbon nanomaterials to adequately assess their toxicity and health risks [438]. Another concern is the need for caution when interpreting the data obtained using these highly complex bio/nanomaterial systems [437]. *Hurt et al.* [438] have emphasized upon the short-term assays and subchronic exposures. They must be conducted in well-characterized in vitro (macrophage and/or epithelial cell cultures) and in vivo (intra tracheal or inhalation in rats, mice or hamsters) models and calibrated using validated negative and positive controls [438]. There is an ongoing debate among toxicologists, industry, and regulatory agencies about the utility of in vitro cellular toxicity assay versus sub-chronic animal testing in screening for potential adverse health effects of nanomaterials [438].*Hurt et al.* [438] discussed the overall status and trends of toxicology of carbon related nanomaterials. They mentioned that instead of supporting either extreme viewpoint – that nanomaterials pose no health risks, or that nanotechnology presents extreme risks that warrant cessation of development activities, it is better to continue research on toxicity and safety with a wide range of new material formulations. A worthy goal for toxicologists and materials scientists is the joint development of 'green' nanomaterial formulations – those co-optimized for function and minimal health impact. Health effects data on workers involved with nanotechnology are limited because of the incipient nature of the field, the relatively small number of workers potentially exposed to date, and the lack of time for chronic disease to develop and be detected. The most relevant human experience deals with exposures to ultrafine particles (which include particles with diameters < 100 nm) and fine particles (particles with diameters < 2.5 μm).Ultra fine and fine particles have been assessed in epidemiologic air pollution studies and in studies of occupational cohorts exposed to mineral dusts, fibers, welding fumes, combustion products, and poorly soluble, low-toxicity particulates such as titanium dioxide and carbon black [439-440]. The hazards of these exposures and exposures to engineered nanoparticles are also identified in animal studies [441-442]. A strong relationship exists between the surface area, oxidative stress, and pro-inflammatory effects of nanoparticles in the lung. The greater the oxidative stress, the more likely the risk of inflammation and cytotoxicity [442]. The findings from animal studies ultimately need to be interpreted in terms of the exposure (dose) that humans might receive. Although there is still some debate, the evidence from air pollution studies associates increased particulate air pollution (the finer particulate matter fraction, PM2.5, with an aerodynamic diameter < 2.5 μm) with adverse health effects in susceptible members of the population—particularly the elderly with respiratory and cardiovascular diseases [443]. Moreover, the concentrations associated with measurable effects on the health of populations are quite low [444]. In occupational studies, the populations that are repeatedly exposed to hazardous mineral dusts and fibers in the respirable range (e.g., quartz and asbestos, respectively) have well-known health effects related to the dose inhaled [444]. With asbestos, the critical risk factors for developing respiratory diseases are fiber length, diameter, and bio-persistence. For poorly soluble, low-toxicity dusts such as titanium dioxide, smaller particles in the nanometer size range appear to cause an increase in risk for lung cancer in animals on the basis of particle size and surface area [445].Although the findings are not conclusive, various studies of engineered nanoparticles in animals raise concerns about the existence and severity of hazards posed to exposed workers [446]. Possible adverse effects include the development of fibrosis and other pulmonary effects after short-term exposure to carbon nanotubes [441-442],the translocation of nanoparticles to the brain via the olfactory nerve, the ability of nanoparticles to translocate into the circulation, and the potential for nanoparticles to activate platelets and enhance vascular thrombosis [447].None of these findings are conclusive about the nature and extent of the hazards, but they may be sufficient to support precautionary action. Ultimately, the significance of hazard information depends on the extent to which workers are exposed to the hazard. This is the defining criterion of risk (the probability that an exposed worker will become ill).There is a need identified for nanoparticle-specific risk assessments (i.e., those that use the most appropriate dosemetrics rather than typical mass) that will be unique to nanotechnology [448].Risk assessment has been widely used to manage the uncertainty of risks posed to humans by newly introduced chemicals or processes. However, nanotechnology encompasses a diverse range of compositions, structures, and applications, so a single risk assessment and management strategy may not be appropriate [449].

13. Ethical and Scientific Issues of Nanotechnology in the Workplace

The ethical issues that most affect workers in jobs involving nanomaterials are linked to identification and communication of hazards and risks by scientists, authorities, and employers; acceptance of risk by workers; implementation of controls; choice of participation in medical screening; and adequate investment in toxicological and exposure control research ,Table.13.

The ethical issues involve the identification and assessment of hazards and risks, nonmaleficence (doing no harm), autonomy (self-determination), justice (fairness in distribution of risks), privacy (in handling of medical information),and respect for persons. Factual scientific knowledge—which is the basis for ethical decisions about occupational safety and health—may be influenced by biases and values [450].Scientific knowledge is unavoidably value laden. No scientific theory can be considered to be wholly objective, but one theory may be more objective than another [451]. Underlying the ethical decisions are the way in which nanotechnology is depicted, the potential benefits, and the associated hazards and risks. When information about the hazards of nanoparticles is in doubt, the critical question is where to draw the line about the necessary level of protection and the residual risk at a given level of protection. Risk assessments are partly subjective and likely to be highly politicized. Thus all risk projections are value laden. No single scenario for describing risks and controls can suffice because of the heterogeneous and developmental nature of nanotechnology. The ethical issues will be specific only for the knowledgebase at a given time and for a specified production and use scenario. Researchers have suggested that even with that type of specificity, alternative assessments are needed to capture the ethical and political values that inform policies such as those involving nanotechnology [451].Employers and workers look to scientists and authoritative organizations to help interpret hazard and risk information and to put it into context. This expectation may pressure scientists to go beyond the mere conduct of research. The interface between science and morality is exceedingly complex, but scientists are generally considered to have ethical obligations to society at large [452].However, no consensus has been reached about the nature of those ethical obligations beyond fulfilling the professional responsibilities internal to scientific research. Framing a clear and coherent approach to the ethical responsibilities of scientists in nanotechnology is a difficult task. At the least, such an approach requires scientists to use appropriate qualifiers in published papers and to be cautious in generalizing their results. More broadly, it means not shrinking from considering the implications of their work, even if all the scientific details are not known. Decision makers may have inadequate scientific information to help them decide how precautionary their approach should be [451]. To determine whether a decision conforms to the principle of normal eficence, decision makers must determine the harm that could occur if the nanoparticles were as toxic as suggested by preliminary hazard information. Data on air pollution and industrial ultra fine particles indicate that a given mass of nanoparticles would be more biologically reactive and hence potentially more toxic than the same mass of larger particles [453]. Consequently, the level of control might need to be more stringent for smaller nanoscale dusts than for those with diameters > 100 nm. Ultimately, the more stringent level of controls may result in risks that are equal to or smaller than risks posed by larger particles. Authoritative organizations and employers are responsible for communicating the risk workers face after appropriate controls are implemented. Failure to do so may preclude workers from exercising autonomy. This issue may be confounded by the fact that the employer has a proprietary interest in not releasing information about "nanoproducts" and workplace controls. The principlist ethical approach focuses on principles such as nonmaleficence and autonomy but fails to assess the social and organizational context of occupational safety and health and the role of practitioners in relation to the corporate structure [454]. With regard to nanotechnology, the contextual pressures on practitioners and authorities arise from a company's or society's needs and desires for nanotechnology to grow and develop. Mention of potential health concerns may be seen as alarmist, unfounded, and detrimental to the growth of the field. Nonetheless, the counter position is that conflicting demands on practitioners from being both an agent of a company and an autonomous professional constitute a social and structural problem rather than a problem of individual ethics [455]. One solution is that health pronouncements be made independently of promotional concerns for nanotechnology. Acceptance of risk is a relative concept that includes judgment about the certainty and severity of risk, the extent of the health effects, voluntary nature of the risk, the risks and advantages of any alternatives, and compensation for undergoing the risk. It is a false premise to assert that workers have free choice in terms of which work and working conditions to accept. Although some component of self-determination is present, economic and social conditions exert the greatest influences on workers selection of work, level of risk tolerated, and ability to participate in risk management. Worker participation in risk management is not a static concept and has increased over the past 35 years with the implementation of team approaches, management systems, corporate responsibility, and right to know and act movements [456-457]. Nonetheless, workers generally cannot universally refuse work they consider hazardous and still keep their jobs. Conformance with the principle of autonomy depends on the extent to which workers have input into risk management at their work sites and the degree to which they are at risk after controls have been implemented. Justice is also related to worker decision making. At issue is the extent to which workers are exposed to greater risks than the general public—or, stated another way, whether it is appropriate to exchange incentives such as wages or hazardous duty pay for additional risk from exposure to nanoparticles [451]. This issue may be less significant if nanoparticle controls reduce workers' risk levels to those of the general public, if conceivably both are known. Clearly, society accepts that some jobs are inherently riskier than others. However, in many countries the societal goal is to provide a safe and healthful workplace for all workers.

14. Epilogue

The potential of nano conducting polymers and nanocomposites in various sectors of research and application is promising and attracting increasing investment from Governments and business in many parts of the world. While there are some niche applications where nanotechnology has penetrated the market, the major impact will be at least a decade away. Currently, there are a few cosmetic products on the market made by incorporating nanoparticles. To create such macro-scale materials, many issues surrounding the incorporation of nanotubes into a matrix, strategies for property improvement, and the mechanisms responsible for those property improvements still remain critical. Since only a moderate success has been made over the last 20 years, researchers must continue to investigate strategies to optimize the fabrication of nanotube-enabled materials. Biodegradable polymer-based nanocomposites have a great deal of future promise for potential applications as high-performance biodegradable materials. These are entirely new types of materials based on plant and nature materials (organoclay). When disposed of in compost, these are safely decomposed into CO_2, water, and humus through the activity of microorganisms. Undoubtedly, their unique properties originating from the controlled nanostructure paves the way for a much broader range of applications (already commercially available through Unitika Ltd., Japan), and opens a new dimension for plastics and composites. According to several sources (Chemical Business News base, Plastic News), a significant increase in turnover of about 100%/year, leading to a value of about 1500 million Euro (ca. 500,000 ton/year of polymer nanocomposites) in 2009 is expected. Even if these numbers are overly optimistic, they highlight the tremendous technological and economical potential associated with polymeric nanocomposites including not only clay but also other inorganic nanofillers, such as carbon nanotubes. At present it is difficult to predict which, if any, market sector would not be able to benefit from this technology. Thus, it may well be that polymeric nanocomposites in the mid and longer-term will pervade all aspects of life, similar to the way plastics did in the last century. Clearly a diverse range of sectors such as aerospace, automotive, packaging (particularly food but also solar cells), electrical and electronic goods, household goods etc. will profit substantially from a new range of materials which would be offered by this technology. In the short term (<5 years), the commercial impact may include inkjet markets, nanoparticles in cosmetics, and automotive applications such as body moldings, engine covers and catalytic converts, batteries, computer chips . In the mid-term (10 years), memory devices, biosensors for diagnostics, advances in lighting are all possible. The time-scale for automotive, aerospace, bio-nanotechnology is a long-term prospect (>15 years) as these are risk-averse sectors and thereby for large-scale production it is necessary to carry out strict testing and validation procedures.

REFERENCES

[1] Zhao, H. and Ning, Y. (2000).Techniques Used for the Preparation and Application of Gold Powder in Ancient China. *Golden Bulletin*, 33(3), 103-105

[2] Faraday, M. (1857).The Bakerian lecture: experimental relations of gold.*Philosophical Transactions,* 147-145

[3] Feynman, R.P. (1961) Miniaturization New York: Reinhold

[4] Ogawa, H., Nishikawa, M. and Abe, A. (1982).Hall measurement. *Journal of Applied Physics*, 53, 4448 - 4449

[5] David, J. R.Topics for Investors © 2004

[6] Zheng, MA. Jagota, E. Semke, B.A. Diner, R.S. McLean, S.R. Lustig, R.E. Richardson and Tassi N.G. (2003). DNA assisted dispersion and separation of carbon nanotubes. *Nature Materials*, 338-342

[7] Skotheim, T.A., Elsenbaumer, R.L., Reynolds, J.R.:(Eds.) (1998) Handbook of Conducting Polymers, 2nd edition, Marcel Dekker: New York

[8] Martin, C.R. (1994). Nanomaterials: A Membrane-Based Synthetic Approach. *Chem. Mater.*, 266, 1961-1966.

[9] Shirakawa, H., Louis, E. J., MacDiarmid ,A.G., Chiang C. K., Heeger A. J.(1977). Synthesis of electrically conducting organic , *J. Chem. Soc .Chem. Commun*. 578-580.

[10] T. J. Skotheim, R. L. Elsenbaumer, J. R. Reynolds. Handbook of Conducting Polymers, 2nd ed.,pg 1097, Marcel Dekker, New York (1998).

[11] Chandrasekhar, P. (1999). Conducting Polymers, Fundamentals and Applications: A Practical Approach, p. 60, Kluwer Academic, Boston

[12] Ranjani, V. Parthasarathy, Charles R. M. (1994) Template-Synthesized Polyaniline Microtubules. *Chem. Mater.*, 6 (10), 1627–1632

[13] Sailor, M.J. and Curtis, C.L. (1994). Conducting polymer connections for molecular devices, *Adv. Mater*. 6(9), 688-692

[14] Neves ,S., Gazotti, W.A., De Paoli M.A. (2004).In Encyclopedia of Nanoscience and Nanotechnology ,Vol. 2, Nalwa H. S. (Ed.), American Scientific Publishers, Los Angeles. pp. 133–152.

[15] Gangopadhyay , R. (2004). In Encyclopedia of Nanoscience and Nanotechnology, Vol. 2, Nalwa H. S. (Ed.), American Scientific Publishers, Los Angeles pp. 105–131

[16] Goldberger, J., He, R., Zhang, Y., Lee, S., Yan, H., Choi, H.J. Yang, P. (2003). Single-crystal gallium nitride nanotubes. *Nature*, 422, 599-602.

[17] Yin, L.-W., Bando, Y., Zhan, J.-H., Li, M.-S., Golberg, D. (2005) .Self-Assembled Highly Faceted Wurtzite-Type ZnS Single-Crystalline Nanotubes with Hexagonal Cross-Sections. *Adv. Mater.*, 17(16), 1972-1977.

[18] Xia, Y.; Yang, P., Sun, Y., Wu, Y., Mayers, B., Gates, B., Yin, Y., Kim, F., Yan, H.(2003). One-dimensional nanostructures. *Adv. Mater*, 15, 353-389.

[19] Brumlik, C .J., Martin, C. R. (1991) Template synthesis *J.Am.Chem.Soc.*,113, 3174-3175.

[20] Johnson, S. A., Ollivier, P. J., Mallouk, T. E. (1999) Ordered Mesoporous Polymers of Tunable Pore Size from Colloidal Silica Templates Science, 283 (5404),963-965.

[21] Kline, T.R., Tian, M., Wang, J., Sen, A., Chan, M.W.H., Mallouk, T. E. (2006). Template-grown metal nanowires, *Inorg.Chem*, 45, 7555-7565.

[22] Shelimov, K.B., Moskovits, M. (2000) Composite Nanostructures Based on Template-Grown Boron Nitride Nanotubules, *Chem. Mater*. 12(1), 250-254.

[23] Hsiao, G., Anderson, M.G., Gorer, S., Harris, D., and Penner, R. M. (1997). Hybrid Electrochemical/Chemical Synthesis of Supported, Luminescent Semiconductor Nanocrystallites, *J. Am. Chem. Soc*. 119, 1439-1448

[24] Klein, J.D., Herrick, R.D.II, Palmer, D., Sailor, M. J., Brumlik C.J., Martin C.R. (1993) Electrochemical Fabrication of Cadmium Chalcogenide Microdiode Arrays, *Chem. Mater*. 5(7), 902–904.

[25] Wallace, G.G., Innis, P.C., Kane-Maguire, L.A.P.(2004).In Encyclopedia of Nanoscience and Nanotechnology, Vol. 4, H. S. Nalwa (Ed.), American Scientific Publishers, Los Angeles. pp. 113–130

[26] Epstein , A.J. (2001).In Organic Electronic Materials: Conjugated Polymers and Low Molecular Weight Organic Solids, Vol. 41, R. Farchioni and G. Grosso (Eds.), p.3, Springer, Amsterdam

[27] Martin, C.R. (1995).Template Synthesis of Electronically Conductive Polymer Nanostructures, *Acc.Chem.Res*. 28(2), 61-68

[28] Zhang, Z., Wei, Z., Zhang, L., Wan, M. (2002) Polyaniline nanotubes and their dendrites doped with different naphthalene sulfonic acids, *Acta Materialia*, 53(5), 1373-1379.

[29] Xiao, R., Cho, S., Liu,R. and Lee, S.B. (2007) Controlled Electrochemical Synthesis of Conductive Polymer Nanotube Structures , *J. Am. Chem. Soc.*, 129, 4483-4489.

[30] Liu, L., Zhao, C., Zhao, Y., Jia, N., Zhou, Q., Yan M., Jiang, Z. (2005). Characteristics of polypyrrole (PPy) nano-tubules made by templated ac electropolymerization, *Eur. Polym. J*, 41(2) , 2117–2121

[31] Mantovani, J. G., Warmack, R. J. Annis, B. K. , Macdiarmid, A.G. , Scherr, E.(1990).Investigation of surface morphology of emeraldine hydrochloride by scanning tunneling microscopy, *J. Appl. Polym. Sci*.40(9-10), 1693 - 1702

[32] Wei, Y. , Sun, Y., Jang, G.-W. , Tang, X. (1990). Effects of p-amino diphenylamine on electrochemical polymerization of aniline, *J.Polym.Sci*. Part C: Polym. Lett. 28(3), 81-87

[33] Choi, S. J. and Park, S. M. (2002). Electrochemistry of Conductive Polymers. XXVI. Effects of Electrolytes and Growth Methods on Polyaniline Morphology, *J. Electrochem. Soc*.149(2), E26-E34

[34] Liu, J.,Lin, Y.H. , Liang, L., Voigt, J. A. , Huber, D. L., Tian, Z. R., Coker, E. , Mckenzie, B. Mcdermott, M.J. (2003). Templateless Assembly of Molecularly Aligned Conductive Polymer Nanowires: A New Approach for Oriented Nanostructures, *Chem. Eur. J.* 9(3), 604-611

[35] Liang, L. , Liu, J., Windisch, C. F. , Exarhos, G. J. , Lin. Y. H. (2002). Direct Assembly of Large Arrays of Oriented Conducting Polymer Nanowires, Angew. Chem., *Int. Ed.* 41(19), 3665-3668

[36] Huang, J. X. and Kaner, R.B. (2004). Nanofiber Formation in the Chemical Polymerization of Aniline: A Mechanistic Study .Angew.Chem., *Int. Ed.* 43(43), 5817-5821

[37] Chien, J.C.W. ,Yamashita, Y., Hirsch, J. A. , Fan, J. L. , Schen, M. A., Karasz, F. E. (1982). Resolution of controversy concerning the morphology of polyacetylene, *Nature* 299, 608-611
[38] Liu, J. M. and Yang, S.C. (1991). Novel colloidal fibrils made by template guided chemical polymerization, *Chem. Commun*. 21, 1529-1531.
[39] Li, W. G. and Wang, H. L. (2004). Oligomer-Assisted Synthesis of Chiral Polyaniline Nanofibers *J. Am. Chem. Soc.* 126(8), 2278-2279
[40] Qiu, H. J., Wan, M. X., Matthews, B., Dai, L. M. (2001). Conducting Polyaniline Nanotubes by Template-Free Polymerization, Macromolecules 34(4), 675-677
[41] Chao,D., Chen, J., Lu, Chen, X.L., Zhang, W. Wei, Y. (2005). SEM, *Synth. Met*. 150(1) , 47-51
[42] Huang, J. X. and Kaner, R.B.(2004). A General Chemical Route to Polyaniline Nano fibers, *J. Am. Chem. Soc*. 126(3) , 851-855
[43] Zhang, X. and Manohar, S.K. (2004). Polyaniline nanofibers, *Chem. Commun*. 20, 2360-2361
[44] Michaelson, J.C. and McEvoy, A.J. (1994). Interfacial polymerization, *Chem. Commun*. 79-80
[45] Liu, J. M. and Yang, S.C.(1991). Novel colloidal polyaniline, *Chem. Commun*. 1529-1530
[46] Huang, L. M., Wang, Z. B. , Wang, H. T., Cheng, X. L. , Mitra, A., Yan, Y. X. (2002). Polyaniline nanowires by electropolymerization from liquid crystalline phases *J. Mater.Chem*. 12, 388-391
[47] Zhang, X., Goux, W. J. Manohar, S. K. (2004). Synthesis of Polyaniline Nanofibers by "Nanofiber Seeding, *J. Am. Chem. Soc*. 126, 4502-4503
[48] Li, W. G. and Wang, H. L. (2004). Oligomer-Assisted Synthesis of Chiral Polyaniline Nanofibers , *J. Am. Chem. Soc*. 126(8), 2278-2279
[49] Wei , Z.X. and Wan, M.X.(2003). Synthesis and characterization of self-doped poly (aniline-co-amino naphthalene sulfonic acid) nanotubes, *J. Appl. Polym. Sci*. 87(8) , 1297-1301
[50] Wei, Z.X., Zhang, Z.M., Wan , M.X.(2002). Formation Mechanism of Self-Assembled Polyaniline Micro/Nanotubes, Langmuir 18(3) , 917-921
[51] Wan, M.X. In Encyclopedia of Nanoscience and Nanotechnology, Vol. 2, H. S. Nalwa (Ed.),pp. 153–169, American Scientific Publishers, Los Angeles (2004).
[52] Qiu H.J. and Wan, M.X. (2001).Synthesis, characterization, and electrical properties of nano structural polyaniline doped with novel sulfonic acids (4-{n-[4-(4-nitrophenylazo) phenyloxy] alkyl} aminobenzene sulfonic acid), *J. Polym. Sci*., Part A: Polym. Chem. 39(20) , 3485-3487

[53] Hulteen, J.C., Martin, C.R. (1997) A general template-based method for the preparation of nanomaterials *J. Mat. Chem.* 7, 1075-1087

[54] Martin, C.R. (1998) Template polymerization of conductive polymer nanostructures. In: Skotheim TA, Elsenbaumer RL, Reynolds JR(eds) Handbook of conducting polymers. Marcel Dekker, New York, p 409

[55] Treuba, M. , Montero, A.L., Rieumont, J. (2004) Pyrrole nanoscaled electropolymerization: Effect of the proton *Electrochim Acta* 49(3), 4341-4349

[56] Cho ,M.S.,Choi, H.J., Ahn, W.S. (2004) Enhanced Electrorheology of Conducting Polyaniline Confined in MCM-41 Channels, Langmuir 20(1), 202-207

[57] Dauginet-De Pra L, Demoustier-Champagne S (2005) Investigation of the electronic structure, *Polymer* 46(5),1583-1594

[58] Zhang,Z., Sui ,J., Zhang,L., Wan,M., Wei,Y., Yu, L. (2005) Synthesis of Polyaniline with a Hollow, Octahedral Morphology by Using a Cuprous Oxide Template, *Adv. Mater.*17(23),2854-2857

[59] Wang J, Wang J, Wang Z, Wang S (2006) Electrocatalytic oxidation, *Synth. Met.* 156(7-8),610-613

[60] Wang, J., Mo, X., Ge, D., Tian, Y., Wang, Z., Wang, S. (2006). Polypyrrole nanostructures, *Synth. Met.* 156(7-8),514-518

[61] Kulesza, P.J., Chojak, M., Miecznikowski, K., Lewera, A., Malik, M.M., Kuhn A. (2002).Polyoxometallates as inorganic *Electrochem. Commun.* 4(6) ,510-515

[62] Steinmetz, J., Kwon, S., Lee, H.J., Abou-Hamad, E., Almaric, R., Goze-Bac, C., Kim, H., Park, Y.W. (2006).Polymerization of conducting polymers, *Chem. Phys. Lett.* 431(1-3),139-144

[63] Wu, T.M., Lin,Y.W. (2006).Doped polyaniline, *Polymer* 47(10),3576-3582

[64] Marinakos, S.M., Anderson, M.F., Ryan, J.A., Martin,L.D., Feldheim, D.L. (2001).Encapsulation, Permeability, and Cellular Uptake Characteristics of Hollow Nanometer-Sized Conductive Polymer Capsules, *J. Phys Chem.* B,105(37) , 8872-8876

[65] Li, .J, Fang, K., Qiu, H., Li, S.P., Mao, W.M., Wu, Q.Y. (2004).Micro-morphology, *Synth. Met.* 145(2-3),191-194

[66] Krysiński,P., Jackowska,K., Mazur,M.,and Tagowska, M.(2000) Functionalisation of monolayer, *Electochim. Acta.*, 46(2-3), 231-237.

[67] Mazur, M., and Krysinski, P. (2001). Covalently Immobilized 1,4-Phenylenediamine on 11-Mercaptoundecanoic Acid-Coated Gold: Effect of Surface-Confined Monomers on the Chemical in Situ Deposition of Polyaniline and Its Derivatives, Langmuir,17,7093-7101.

[68] Tagowska, M., Mazur, M., and Krysiński, P. (2004).Covalently and ionically immobilized monomers, *Synth. Met.*, 140(1), 29-35.

[69] Sarno, D.M., Manohar, S.K., MacDiarmid, A.G. (2005).Controlled interconversion of semiconducting and metallic forms of polyaniline, *Synth. Met.*, 148(3), 237-243.

[70] Jeevananda, T., Lee, J.H., Siddaramaiah, (2008).Preparation, *Mater. Lett.*, 62(24), 3995-3998.

[71] Chen, F., Xu, Guo-Qin and Andy,T.S.H. (2003). Preparation and assembly of colloidal gold nanoparticles in CTAB-stabilized reverse microemulsion, *Mater. Lett.* 57(21), 3282-3286

[72] Wang, X., Shao, M., Shao, G.; Wu, Z., Wang, S. (2009). A facile route to ultra-long polyaniline, *J. Colloid. Interface. Sci.*,332(1),74-77.

[73] G.Li,Wang,Z., Xie,G., Peng,H., Zhang, Z. (2006). Synthesis of Interconnected Polyaniline Nanofibers in Catanionic Micelles, *J. Dispersion. Sci. Technol.* 27(7), 991-995.

[74] Miao, Z., Wang, Y., Liu, Z., Huang, J., Han, B., Sun, Z., Du, J. (2006). Synthesis of Polyaniline Nanofibrous Networks with the Aid of an Amphiphilic Ionic Liquid, *J. Nanosci. Nanotechn.,* 6(12), 227-230.

[75] Wei, Z., Zhang, Z., Wan, M. (2002). Formation Mechanism of Self Langmuir, 18(3), 917-918.

[76] do Nascimento,G.M., Silva, C.H.B., Izumi, C.M.S., Temperini, M.L.A. (2008) The role of cross-linking, Spectrochim Acta Part A 71(3), 869-875.

[77] Zhang, L., Peng, H., Kilmartin, P. A., Soeller, C., Travas-Sejdic, J. (2008). Poly (3, 4-ethylenedioxythiophene) and Polyaniline Bilayer Nanostructures, Macromolecules 41(20), 7671-7678.

[78] .Zhang, L., Wan, M., Wei, Y. (2006). Nanoscaled Polyaniline Fibers Prepared by Ferric Chloride as an Oxidant, *Macromol. Rapid. Commun.* 27,366-371.

[79] Li, G., Pang, S., Peng, H., Wang, Z., Cui, Z., Zhang, Z. (2005).Templateless and surfactant less route to the synthesis of polyaniline nanofibers, *J. Polym. Sci A: Polym. Chem.* 43(17), 4012-4015

[80] Xia, H., Narayanan, J., Cheng, D., Xiao, C., Liu, X., Chan, H. S. O. (2005) Formation of Ordered Arrays of Oriented Polyaniline Nanoparticle Nanorods, *J. Phys. Chem.* B, 109(26), 12677-12684.

[81] Anilkumar, P., Jayakannan, M.(2006).New Renewable Resource Amphiphilic Molecular Design for Size-Controlled and Highly Ordered Polyaniline Nanofibers Langmuir, 22(13), 5952-5957.

[82] Cheng, C., Jiang, J., Tang, R., Xi, F. (2004).Polyaniline nanostructures , *Synth. Met.*145 (1), 61-65.

[83] Yang, C-H , Chih, Y-K, Cheng, H-E and Chen, C-H. (2005). Nanofibers of self-doped polyaniline Polymer, 46(24),10688-10698.

[84] Mi, H., Zhang, X., Yang, S., Ye, X., Luo. (2008) Polyaniline nanofibers, *J. Mater. Chem. Phys.*,112(1), 127-131.

[85] Xing, S., Zheng, H.(2008) Preparation of polyaniline nanofibers using the organic solution of aniline as seed , e-Polymers,106,1-9.

[86] Thiyagrajan, M., Samuelson, L., Kumar, J., Cholli, A. L.(2003). Helical Conformational Specificity of Enzymatically Synthesized Water-Soluble Conducting Polyaniline Nanocomposites, *J. Am. Chem. Soc.* 125(38), 11502-11503

[87] Thiyagarajan, M., Kumar, J. Samuelson, L., Cholli, A.L. (2003). Enzymatically Synthesized Conducting Polyaniline Nanocomposites: A Solid-State NMR Study , *J. Macromol. Sci. Pure. Appl. Chem.* A 40(12),1347-1355

[88] Ryu, K., Mc Eldoon, J. P., Pokora, A. R. , Cyrus, W. , Dordick. J. S. (1993). Numerical and Monte Carlo simulations of phenolic polymerizations catalyzed by peroxidase, *Biotechnol. Bioeng.* 42(7), 807-814

[89] Akkara, J.A., Senecal, K. J. , Kaplan, D.L. (1991). Synthesis and characterization of polymers produced by horseradish peroxidase in dioxane, *J. Polym. Sci., Polym. Chem*, 29(11), 1561-1574

[90] Kobayashi, S., Uyama, H., Kimura. (2001).Enzymatic Polymerization, *A. Chem. Rev.* 101(12), 3793-3818

[91] Samuelson, L.A., Anagnostopoulos, A., Alva, K.S., Kumar, J., Tripathy, S. K. (1999).Biologically Derived Conducting and Water Soluble Polyaniline Macromolecules, 31(13),4376-4378

[92] Liu, W., Kumar, J., Tripathy, S. , Senecal, K.J. , Samuelson. L. (1999).Enzymatically Synthesized Conducting Polyaniline. *J. Am. Chem. Soc.* 121(1), 71-76

[93] Liu, W., Cholli, A.L., Nagarajan, Kumar, R.J., Tripathy, S.F., Bruno, F., Samuelson. L. (1999) The Role of Template in the Enzymatic Synthesis of Conducting Polyaniline, *J. Am. Chem. Soc.* 121(49) , 11345-11355

[94] Dunford, H.B. In Peroxidases in Chemistry and Biology.(1991). Everse, J. Everse, K.E., Grisham, M.B. (Eds.), CRC Press, Boca Raton, FL, 2, 24

[95] Xu, P., Kumar, J., Samuelson, L. , Cholli, A.L. (2002). Monitoring the Enzymatic Polymerization of 4-Phenylphenol by Matrix-Assisted Laser Desorption Ionization Time-of-Flight Mass Spectrometry: A Novel Approach, Biomacromolecules, 3(5) 889-893

[96] Sahoo, S. K., Liu, W., Samuelson, L., Kumar, J. , Cholli , A. L. In NMR Spectroscopy of Polymers in Solution and Solid State. (2002). Cheng, H.N.

and English , A.D. (Eds.), pp. 258–269, American Chemical Society, Washington, DC

[97] Im , S.G. , Yoo , P.J. , Hammond , P.T. , Gleason ' K. K. (2004) Grafted Conducting Polymer Films for Nano-patterning onto Various Organic and Inorganic Substrates by Oxidative Chemical Vapor Deposition, *Adv. Mater.*, 19(19), 2863 – 2867.

[98] Subrahmanya, M.S.S.(2006), Doctoral thesis, 07 December .

[99] Xing,, S., Jing,S., Zhao, C., Wang, Z. (2007)Effect of addition of organic solvents on properties of the chemically synthesized polyaniline in aqueous solutions e-Polymers, 44,1-10

[100] Zhang , L.J., Wan, M.X. (2003). Self-Assembly of Polyaniline - From Nanotubes to Hollow Microspheres, *Adv. Funct. Mater.,*13(10) , 815-820 .

[101] Zhang, L.J., Long, Y.Z., Chen, Z.J., Wan, M.X. (2004), The Effect of Hydrogen Bonding on Self-Assembled Polyaniline Nanostructures, *Adv. Funct. Mater.*14 (7), 693-698.

[102] Xing , S.X., Zhao ,C., Jing ,S.Y., Wang ,Z.C.(2006).Morphology and conductivity Polymer, 47(7), 2305-2313.

[103] Liu, Y.C. (2003).Evaluation of the Conductivity of the Polypyrrole Film via Its C-C Bonds Stretching on Surface-Enhanced Raman Spectrum, Electroanalysis,15(13), 1134-1138.

[104] Hassan, P.A., Sawant, S. N. Bagkar, N.C. and Yakhmi, J.V. (2004) Polyaniline Nanoparticles Prepared in Rod like Micelles Langmuir, 20, 4874-4880.

[105] Rao, P.S., Subrahmanya, S. and Sathyanarayana, D.N. (2002). Inverse emulsion polymerization, *Synth. Met.*, 128(3), 311-316.

[106] Yin, W. and Ruckenstein, E. (2000).Soluble polyaniline , *Synth. Met.,*108(1), 39-46

[107] Svelko, N.K., Reynaud, S. and Francois, J. (2005). Synthesis and characterization of polyaniline, *Synth. Met.*, 150(2), 107-114.

[108] Huo, L.H., Cao, L.X., Wang, D.M. Cui, H. N. Zeng, G.F. and Xi S.Q. (1999). Preparation and characterization of doped polyaniline films, *Thin Solid Films* ,350(1-2),5-9.

[109] De Souza Jr, F.G., Anzai, T.K., Melo, Jr., P.A., Soares, B.G. , Pinto, M.N., Carlos J.(2008). Influence of reaction media on pressure sensitivity of polyanilines doped with DBSA, *J. Appl. Polym. Sci.*,107(4), 2404-2413

[110] Bredas, J.L.,Street,G.B.(1985). Polarons, Bipolarons, and Solitons in Conducting Polymers, *Acc. Chem. Res.* 1985,18, 309-315

[111] Ansari, R. (2006). Polypyrrole Conducting Electroactive Polymers: Synthesis and Stability Studies, *E-Journal of Chemistry* , 3(13), 186-201.

[112] Senevirathne, M.S., Nanayakkara, A.and Senadeera , G. K. R. (2007). A theoretical investigation of band gaps of conducing polymers - polythiophene, polypyrrole and polyfuran, Proceedings of the Technical Sessions, 23 47-53

[113] Guillet, J., (1985) in "Photophysics and Photochemistry", Cambridge University Press, London.

[114] De Mello, J.C., Wittmann, H.F., Friend, R.H. (1997). An improved experimental determination of external photoluminescence quantum efficiency, *Adv. Mater.*, 9(3), 230-232.

[115] Palsson, L.O., Monkman, A.P. (2002).Measurements of Solid-State Photoluminescence Quantum Yields of Films Using a Fluorimeter, *Adv. Mater.* 14(10), 757-758.

[116] Zory, P.S., Reisinger, A.R., Waters, R.G., Mawst, L.J., Zmudzinski, C.A., Emanuel, M.A., Givens, M.E. and Coleman, J.J. (1986). Anomalous temperature dependence of threshold for thin quantum well AlGaAs diode lasers, *Appl. Phys. Lett.*,49(1), 16-18.

[117] Chosrovian, H., Rentsch, S., Grebner, D., Dahm, D.U. , Birckner, E., Naarmann, H. (1993). Time-resolved fluorescence, *Synth. Met.*, 60(1), 23-26.

[118] Migashita, K., Kaneto, M.(1994) Poly(p-Phenylene) thin films prepared by vacuum deposition and their photo luminescent characteristics, *Macromol. Chem. Rapid Commun.*15 (6), 511-516.

[119] Schmidt, S. , Anderson, M. L. , Dunphy, D., Wehrmeister, T. , Müllen, K., Armstrong, N.R. (1995). Photoelectron and optical spectroscopic investigations of the electronic structure of oligo (p-phenylenevinylene) s in the solid state, *Adv. Mater.*, 7(8), 722-726.

[120] Klessinger, M., Michl, J. (1995).Excited States and Photochemistry of Organic Molecules; New York: VCH.

[121] Chen ,K.H., Yang, S.M. (2003) Polyaniline-Montmorillonite Composite Synthesized by Electrochemical Method, *Synth. Met.*, 135-136,151-152

[122] do Nascimento, G.M., Constantine, V.R.L., Landers, R., Temperini M.l.A. (2004) Aniline Polymerization into Montmorillonite Clay: A Spectroscopic Investigation of the Intercalated Conducting Polymer Macromolecules, 37 (25), 9373–9385

[123] Kim, B.H., Hong, J.H., Hong, S.H., Kim, J.W., Choi, H.J., Joo, J. (2001) Physical characterization of emulsion intercalated polyaniline, *Curr. Appl. Phys.* 1(1), 112-115

[124] Cho, M.S., Cho, Y. H., Choi, H.J.and Jhon M.S.(2003) Synthesis and Electrorheological Characteristics of Polyaniline-Coated Poly(methyl methacrylate) Microsphere: Size Effect, Langmuir, , 19 (14), 5875–5881

[125] Alexandre, M., Dubois, P. (2000). Polymer-layered silicate, *Mater. Sci. Eng. Reports*, 28(1-2), 1-63.

[126] Kawasumi, M.J. (2004). The discovery of polymer-clay hybrids *Polym. Sci., Part A: Polym. Chem.*, 42(4), 819-824.

[127] Weimer, M. W., Chen, H.,Giannelis, E. P.,Sogah, D. Y. (1999). Direct Synthesis of Dispersed Nanocomposites, *J. Am. Chem. Soc.* 121(7), 1615-1616.

[128] Manias, E., Touny, A., Wu, L., Strawhecker, K., Lu, B., Chung, T. (2001). Polypropylene/Montmorillonite Nanocomposites *Chem. Mater.* 13 (10), 3516-3523.

[129] Di, J., Sogah, D.Y. (2006). Exfoliated Block Copolymer/Silicate Nanocomposites, Macromolecules,39(15), 5052-5057.

[130] Shi, Y., Peterson, S., Sogah, D.Y. (2007). Surfactant-Free Method for the Synthesis of Poly (vinyl acetate, *Chem. Mater.* 19(7), 1552-1564.

[131] Haraguchi, K., Takehisa, T. (2002).Nanocomposite Hydrogels: A Unique Organic-Inorganic Network Structure with Extraordinary Mechanical, Optical, and Swelling/De-swelling Properties *Adv. Mater.* 14(16), 1120-1124.

[132] Nazzal, A.I., Street, G.B. (1985) Pyrrole-styrene graft copolymers, *J. Chem. Soc. Chem. Commun*, 375–376.

[133] Sun, Y., Ruckenstein, E. (1995). Polypyrrole-bearing conductive composite prepared by an inverted emulsion pathway involving nonionic surfactants, *Synth. Met.* 72, 261–267.

[134] Wang, L., Brazis, P., Rocci, M., Kannewurf, C.R., Kanatzidis, M.G.(1998).A new redox host for intercalative polymerization: insertion of polyaniline into a-$RuCl_3$. *Chem. Mater*; 10:3298–3300.

[135] Sinha , R.S., Biswas ,M. (1999). Preparation and evaluation of composites from montmorillonite and some heterocyclic polymers: 3. A water dispersible nanocomposite from pyrrole–montmorillonite polymerization system. *Mater. Res. Bull.* 35, 1187–1194.

[136] Kim, B.H., Jung, J.H., Joo, J., Kim, J.W., Choi, H.J. (2001). Nanocomposite of polyaniline and Na-montmorillonite (MMT) clay. Nanocomposites, Proceedings. Chicago, Illinois, USA: ECM Publication.

[137] Kim, J.W., Liu, F., Choi, H.J., Hong, S.H., Joo, J. (2003) Intercalated polypyrrole/Na-montmorillonite nanocomposite via an inverted emulsion pathway method. *Polymer*, 44, 289–293.

[138] Biswas, M., Sinha, R.S. (1998). Preparation and evaluation of composites from montmorillonite and some heterocyclic polymers. 1: poly (N-vinylcarbazole)–montmorillonite system. *Polymer*, 39, 6423–6438.

[139] Sinha, R.S., Biswas, M. (1999) Preparation and evaluation of composites from montmorillonite and some heterocyclic polymers, II. Nanocomposite from N-vinylcarbazole and ferric chloride impregnated montmorillonite polymerization system. *J. Appl. Polym. Sci.*, 73, 2971–2976.

[140] Kim, J.W., Kim, S.G., Choi, H.J., Jhon, M.S. (1999). Synthesis and electro-rheological properties of polyaniline–Na-montmorillonite suspensions. *Macromol. Rapid. Commun.* 20, 450–452.

[141] Biswas, M., Sinha, R.S. (2000). Water-dispersible nanocomposites of polyaniline and montmorillonite, *J. Appl. Polym. Sci.*, 77, 2948–2956.

[142] Dai, L., Wang, Q., Wan, M. (2000).Direct observation of conformational transitions for polyaniline chains intercalated in clay particles upon secondary doping. *J. Mater. Sci. Lett.*, 19, 1645–1647.

[143] Lee, D., Lee SH, Char K, Kim J. (2000). Expansion distribution of basal spacing of the silicate layers in polyaniline/Na- montmorillonite nanocomposites monitored with X-ray diffraction, *Macromol. Rapid. Commun.* 21, 1136–1139.

[144] Wu, Q., Xue, Z., Qi, Z., Wang, F. (2000) .Synthesis and characterization of PANI/clay nanocomposite with extended chain conformation of polyaniline. *Polymer*, 41, 2029–2032.

[145] Uemura, S., Yoshie, M., Kobyashi, N., Nakahira, T. (2000). Photo polymerization of aniline dimer by photocatalytic reaction of ruthenium trisbipyridyl in the interlayer of hectorite clay, *Polym. J*, 32, 987–990.

[146] Kim , B.H., Jung, J.H., Joo, J., Kim, J.W., Choi ,H.J.(2000) Charge transport and structure of nanocomposites of polyaniline and inorganic clay, *J. Korean Phys. Soc.*, 36, 366–370.

[147] Yeh, J.M., Liou, S.J., Lai, C.Y., Wu, P.C., Tsai, T.Y. (2001). Enhancement of corrosion protection effect in polyaniline via the formation of polyaniline–clay nanocomposite materials, *Chem. Mater.*, 13, 1131–1136.

[148] Feng, B., Su, Y., Song, J., Kong, K. (2001). Electropolymerization of polyaniline/montmorillonite nanocomposite, *J. Mater. Sci. Lett.*, 20,293–294.

[149] Kim, J.W., Kim, S.G., Choi, H.J., Suh, M.S., Shin, M.J., Jhon, M.S. (2001). Synthesis and electrorheological characterization of polyaniline and Na-montmorillonite clay nanocomposite. *Int. J. Mod. Phys.,* 15,657–664.

[150] Kim, B.H., Jung, J.H., Kim, J.W., Choi, H.J., Joo, J. (2001) Physical characterization of polyaniline–Na-montmorillonite nanocomposite intercalated by emulsion polymerization, *Synth. Met.* , 117,115–118.

[151] Choi, H.J., Kim, J.W., Joo, J., Kim, B.H. (2001). Synthesis and electrorheology of emulsion intercalated PANI–clay nanocomposite, *Synth. Met.*, 121,1325–1326.

[152] Cho, M.S., Choi, H.J., Kim ,K.Y., Ahn, W.S. (2002).Synthesis and characterization of polyaniline/mesoporous SBA-15 nanocomposite, *Macromol .Rapid. Commun.*,23, 713–716.

[153] Kim, B.H., Jung ,J.H. , Hong, S.H., Joo, J., Epstein,A.J., Mizoguchi , K., Kim ,J.W.,Choi ,H.J.(2002), Nanocomposite of polyaniline and Na-montmorillonite clay, Macromolecules, 35:1419–1423.

[154] do Nascimento, G.M., Constantino ,V.R.L, Temperini M.L.A. (2002) Spectroscopic characterization of a new type of conducting polymer–clay nanocomposite, Macromolecules,35, 7535–7537.

[155] Winkler, B., Dai, L., Mau ,A.W-H. (1999). Organic–inorganic hybrid light-emitting composites: poly (p-phenylene vinylene) intercalated clay nanoparticles. , *J. Mater. Sci. Lett.*, 18, 1539–1541.

[156] Ke, S., Ying, M., Yaan, C., Zhao-Hui, C., Xue-Hai , J., Jianniah ,Y. (2001). Inclusion of poly (tetra methyl-p-phenylenediamine dihydrochloride) into MoO3: a cooperative formation route to construct a polymer/MoO_3 layered structure. *Chem. Mater.*, 13, 250–252.

[157] Pennwell, R.C., Ganguly, B.N., Smith, T.W. (1973). Poly (N-vinyl carbazole): a selective review of its polymerization, structure, properties, and electrical characteristics. *J. Polym. Sci, Macromol. Rev.*, 13, 63-160.

[158] Biswas, M., Das, S.K.(1982) .Recent progress in carbazole based polymers. *Polymer,* 23, 1713–1726.

[159] Biswas, M., Sinha, R.S. (1998) A poly (N-vinylcarbazole) based nanocomposite form montmorillonite. In: Srinivasan KSV, editor. Macromolecules new frontiers. Proceedings of the IUPAC International Symposium on Advances in Polymer Science and Technology, MACRO'98, vol. II, New Delhi: Allied Publishers Ltd;. Pg 710–713.

[160] Biswas, M. , Mitra, P. (1991). Synthesis and thermal stability, dielectric, and conductivity characteristics of some aromatic anhydride-modified carbazole polymers, *J. Appl. Polym. Sci.* 42, 1989–1995.

[161] Inoue, H., Yoneyama, H.J. (1987) Electropolymerization of aniline, *Electroanal. Chem.*, 233(1-2), 291-294

[162] Mehrotra , V., Giannelis, E.P .(1991) Metal-insulator molecular multilayers of electroactive polymers, Solid State Commun. 77(2), 155-158.

[163] Porter, T.L., Eastman, M.P., Zhang, D.Y. and Hagerman, M.E. (1997). Surface Polymerization of Organic Monomers on Cu(II)-Exchanged Hectorite, *J. Phys. Chem.* B, 101(51), 11106-11111

[164] Porter, T.L., Manygoats ,K., Bradley , M., Eastman, M.P., Reynolds,B.P., Votava, A.E., Hagerman, M.E. (1998)Surface and intergallery catalytic properties of Cu(II)-exchanged hectorite: A scanning force microscope study, *J. Vac. Sci .Technol.* A, 16(3), 926-931

[165] Eastman, M.P., Hagerman, M.E., Attuso, J.L., Bain, E.D., Porter, T.L. (1996). Polymerization of benzene and aniline on Cu(II)-exchanged hectorite clay films; a scanning force microscope study, Clays. Clay. Mater., 44(6), 769-773

[166] Bae, W.J., Kim, K.H., Jo, W.H., Park, Y.H. (2004).Exfoliated Nanocomposite from Polyaniline Graft Copolymer/Clay, Macromolecules 37(26), 9850-9854

[167] Anuar, K., Murali, S., Fariz, A., Ekramul, H.N.M.M.(2004) Conducting Polymer / Clay Composites: Preparation and Characterization, *Mater. Sci.*,10(3), 255-258

[168] Zarbin, A.J.B., De Paoli, M-A., Alves, O.L. (1997). New polyaniline, *Synth.* Met. 84(1-3) ,107-108

[169] Zarbin, A.J.B., De Paoli, M-A, Alves ,O.L. (1999). Synth. Met. Nanocomposites, 99(3), 227-235

[170] Gill, M., Baines, F.L., Armes, S.P. (1993). Some observations, *Synth. Met.* 55(2-3),1029-1033

[171] Flitton, R., Johal, J., Maeda, S., Armes, S.P. (1995) Synthesis of Colloidal Dispersions of Polypyrrole-Silica Nanocomposites, *J. Colloid. Interf .Sci.,* 173(1),135-142

[172] Butterworth, M.D., Corradi, R., Johal, J., Lascelles, S.F., Maeda, S., Armes, S.P. (1995). Zeta Potential Measurements on Conducting Polymer-Inorganic Oxide Nanocomposite Particles, *J. Colloid. Interf. Sci.*, 174(2), 510-517

[173] Maeda ,S., Armes, S.P. (1993) .Preparation, *J. Colloid. Interf. Sci.*,159(1),257-259

[174] Maeda, S., Armes, S.P. (1995) .Surface area measurements on conducting polymer, *Synth. Mater.* 73(2),151-155

[175] Neoh ,K.G., Tan ,K.K., Goh, P.L., Huang ,S.W., Kang, E.T., Tan, K.L. (1999) Electroactive polymer, *Polymer*, 40(4), 887-893.

[176] Nastase, F., Stamatin, I., Nastase, C., Mihaiescu, D., Moldovan,A .(2006). Synthesis and characterization of PANI, *Prog. Solid. State. Chem.*, 34(2-4), 191-199

[177] Li, X., Wang, G., Li, X. (2005). Surface modification of nano-SiO_2, *Surf. Coat .Technol.*197(1),56-60

[178] Stejskal, J., Kratochvil, P., Armes, S.P., Lascelles, S.F., Riede, A., Helmeted, M., Prokes, J., Krivka, I. (1996). Polyaniline Dispersions. 6.† Stabilization by Colloidal Silica Particles, Macromolecules, 29(21),6814-6816

[179] Kuramoto, N., Takahashi, Y., Nagai, K., Koyama, K. (1996). Electrorheological properties of poly (o-anisidine) and poly (o-anisidine)-coated silica, *React. Funct. Polym.*, 30(1-3),367-373

[180] Gill ,M.T., Chapman ,S.E., De Armitt ,C.L., Baines ,F.L., Dadswell, C.M., Stamper, J.G., Lawless, G.A., Billingham, N.C., Armes, S.P. (1998). A study of the kinetics, *Synth. Met.*, 93(3), 227-233

[181] Liu, P., Liu, W., Xue, Q. (2004). In situ chemical oxidative, *Mater. Chem. Phys*. 87(1),109-113

[182] Li ,Z.F., Swihart ,M.T., Ruckenstein, E. (2004). Luminescent Silicon Nanoparticles Capped by Conductive Polyaniline through the Self, Langmuir, 20(5),1963-1971

[183] Manríquez, V., Galámez, A., Ponce ,J., Brito, I., Kasaneva, J. (1999). Intercalation of polyaniline, *Mater. Res. Bull.*, 34(1), 123-130

[184] Zhang, D., Qin, J., Yakushi, K., Najazawa, Y. (2000). Preparation, *Mater. Sci. Eng.* A ., 286(1), 183-187

[185] Posudievsky, O.Y., Biskulova, S.A., Pokhodenko, V.D. (2002). New polyaniline *J. Mater. Chem.*, 12, 1446-1449

[186] Ballav, N., Biswas, M. (2006). Conductive composites , *Mater. Lett.*, 60(4),514-517

[187] Hill,P.G., Foot , Peter, .J.S., Davis, R. (1996) Novel inorganic/conjugated polymer nanocomposites, *Synth. Met.*, 76(1-3), 289-292.

[188] Parvatikar, N., Jain, S., Khasim, S., Revansiddappa, M., Bloraskar, S.V., Ambika, P.M.V.N. (2006). Electrical and humidity, *Sens. Actuators*. B, 114(2), 599-603

[189] Bhuvanesh, N.S.P., Gopalakrishnan, J. (1998). Polymerization of aniline, *Mater. Sci. Eng.* B, 53(3), 267-271

[190] Kulesza, P.J., Chojak M., Miecznikowski, K., Lewera, A., Malik, M.A., Kuhn, A. (2002) .Polyoxometallates as inorganic, *Electrochem. Commun.* 4(6) ,510-515

[191] Prasad, G.K., Takei, T., Yonesaki, Y., Kumada, N., Kinomura, N. (2006). Hybrid nanocomposite based on $NbWO_6$ nanosheets, *Mater. Lett.* 60(29-30), 3727-3730

[192] Nam, H.J., Kim ,H., Chang, S.H., Kang ,S.G., Byeon, S.H. (1999) Polymerization of aniline , Solid State Ionics, 120(1-4), 189-195

[193] Koenem, B.E., Nazar, L.F. (1996) Synthesis and electrochemical lithium, Solid State Ionics ,89(1-2) ,147-157

[194] Uma, S., Gopalakrishnan, J. (1995). Polymerization of aniline, *Mater. Sci. Eng*. B ., 34(2-3), 175-179

[195] Maia, D.J., Alves, O.L., Paoli, M-A (1997).Growth of linear , *Synt. Met.* , 90(1), 37-40

[196] Zarbin, A.J.G., Maia, D.J., De Paoli, M.A., Alves, O.L. (1999). Polyaniline intercalation *Synth. Met* .102(1-3),1277-1278

[197] Dhamodharan ,R., Jeyaprakash, J. D., Samuel, S., Rajeswari, M. K. (2001) Intercalative redox polymerization and characterization of poly(4-vinylpyridine)-vermiculite nanocomposite, *J. Appl. Polym. Sci.*, 82(3),555-561

[198] Del ,Castillo-Castro T., Larios-Rodriguez, E., Molina-Arenas Z, Castillo-Ortega M.M., Tanori ,J. (2007) Synthesis and characterization of metallic nanoparticles, Composites A .*Appl. Sci. Manu.* 38(1),107-113

[199] Lin ,J., Zhou, W., O'Connor, C.J .(2001). Formation of ordered arrays of gold, *Mater. Lett.* 49(5),282-286

[200] Ullah H., Chung, Won-Sub., Kim, Prof.' Il., Chang-Sik H., Prof. . (2006) pH-Selective Synthesis of Monodisperse Nanoparticles and 3D Dendritic Nanoclusters of CTAB-Stabilized Platinum for Electrocatalytic O_2 Reduction, Small, 2(7),870-873

[201] Smith, J.A., Josowicz, M. , Janata, J.(2003).Polyaniline-Gold nanocomposite system, *J. Electrochem. Soc.*,150(3),:E384-E-388

[202] Zou, X. , Bao, H., Guo, H., Zhang, L., Qi, L., Jiang, J., Niu, L., Dong, S.(2006) Mercaptoethane sulfonate protected, water, *J. Colloid. Interf. Sci.*, 295(2), 401-408

[203] Burst, M., Flink, J., Bethell, D., Schiffrin, D.J., Kiely, C.J. (1995). Synthesis and reactions of functionalized gold, *J. Chem. Soc. Chem. Comm.*, 16, 1655-1656

[204] Chen,S., Kimura, K.(1999).Synthesis and Characterization of Carboxylate-Modified Gold Nanoparticle Powders Dispersible in Water, Langmuir 15(4), 1075-1082

[205] Cheung,J.H. , Stockton, W.B., Rubner, M.F. (1997). Molecular-Level Processing of Conjugated Polymers. 3. Layer-by-Layer Manipulation of Polyaniline via Electrostatic Interactions, Macromolecules 30(9), 2712

[206] Kang ,Y.O., Choi, S.H., Gopalan, A., Lee, K.P., Kang, H.D., Song, Y.S. (2006) Tuning of morphology, *J. Non-Cryst. Solids* 352(5) , 463-468

[207] Pillalamarri, S.K., Blum, F.D., Tokuhiro, A.T., Bertino, M.F. (2005).One-Pot Synthesis of Polyaniline−Metal Nanocomposites, *Chem. Mater.* 17(24), 5941-5944

[208] Khanna, P.K., Singh, N., Charan, S., Viswanath, A.K. (2005). Synthesis of Ag *Mater. Chem .Phys.*, 92(1), 214-219

[209] Zhou, H.H., Ning, X.H., Li ,S.L., Chen, J.H., Kuang, Y.F. (2006).Synthesis of polyaniline *Thin. Solid. Films* 510(1-2) ,164-168

[210] Sivakumar, M., Gedanken, A. (2005). A sonochemical method for the synthesis, *Synth. Met.*148(3), 301-306

[211] Xian, Y., Hu, Y., Liu, F., Xian, Y., Wang, H., Jin, J.(2006). Glucose biosensor based on Au nanoparticles, *Biosens. Bioelectron.*, 21(10) :1996-2000

[212] Ma ,Y., Li ,N., Yang, C., Yang, X.(2005). One-step synthesis, *Colloids. Surf.* A., 269(1-3):1-6

[213] Zheng, L., Li, J.(2005).Platinum–polyaniline, *J. Electroanal. Chem.* , 577(1), 137-144

[214] Huang,S.W. , Neoh,K.G., Shih, C.W., Lim, D.S., Kang, E.T., Han, H.S., Tan, K.L. (1998) Synthesis, characterization and catalytic properties *Synth. Met.* 96(2), 117-122

[215] Ting,Y.P., Neoh, K.G., Kang ,E.T., Tan, K.L. (1994).Recovery of gold by electro less precipitation from acid solutions using polyaniline, *J. Chem. Technol. Biot.*, 59(1), 31-36

[216] Kang, E.T., Ting, Y.P., Neoh, K.G., Tan, K.L. (1995) Electro less recovery, *Synth. Met.* 69(103), 477-478

[217] Houdayer, A., Schneider, R., Billaud, D., Ghanbaja, J., Lambert, J. (2005).Heck and Suzuki-Miyaura couplings catalyzed by nanosized palladium in polyaniline, *Appl. Organometal. Chem.* 19(12) , 1239-1248

[218] Athawale, A.A., Bhagwat, S.V., Katre, P.P. (2006). Nanocomposite of Pd–polyaniline, *Sens. Actuators*. B 114 (1), 263-267

[219] Houdayer, A., Schneider, R., Billaud, D., Ghanbaja, J., Lambert, J. (2005). New polyaniline *Synth. Met.* 151(2),165-174

[220] Houdayer,A., Schneider, R., Billaud ,D., Lambert ,J., Ghanbaja, J. (2007) Preparation, *Mater. Lett.,* 61(1),171-176

[221] Athawale, A.A., Bhagwat, S.V. (2003). Synthesis and characterization of novel copper/polyaniline nanocomposite and application as a catalyst in the Wacker oxidation reaction, *J. Appl. Polym. Sci.,* 89(9), 2412-2417

[222] Sharma, S., Nirkhe, C., Pethkar, S., Athawale, A.A. (2002). Chloroform vapour sensor based on copper, *Sens. Actuators*. B 85(1-2), 131-136

[223] Kanatzidis, M.G., Wu, C.G., Marcy, H.O., Kanewurf, C.R. (1989). Conductive-polymer, *J. Am. Chem. Soc.* 111(11) , 4139-4141

[224] Wu, C.G., De Groot, D.C., Maroy, H.O., Schindler, J.L., Kannewurf, C.R., Liu, Y.J., Horpo, W., Kanatzidis, M.G. (1996) Redox, *Chem. Mater.* 8,1992-2004

[225] Wu, C.G., Liu, Y.C., Hsu, S.S. (1999). Assembly of conducting polymer ,Synth.Met.102(103), 1268-1269

[226] Wu, C.G., Hwang, J.Y., Hsu, S.S. (2001). Synthesis and characterization of processible conducting polyaniline*J. Mater. Chem.* 11, 2061-2066

[227] Petkov, V., Parvanov, V., Trikalitis, P., Malliakas, C., Vogt, T., Kanatzidis, M.G. (2005). Three-Dimensional Structure of Nanocomposites, *J .Am. Chem. Soc.* 127(24) , 8805

[228] De, S., Dey, A., De, S.K. (2005). Charge transport mechanism of vanadium pentoxide xerogel-polyaniline nanocomposite, *Eur. Phys. J.* B. 46(3), 355-362

[229] Huguenin,F., Ferreira ,M., Zucolotto,V., Nart ,F.C., Torresi, R.M., Oliveira ,Jr (2004) Molecular-Level Manipulation of V_2O_5/Polyaniline Layer-by-Layer Films To Control Electrochromogenic and Electrochemical Properties, *Chem. Mater.* 16(11) , 2293-2299

[230] Ferreira,F., Zucolotto, V., Huguenin, F., Torres, R.M., Oliveira ,Jr.,.(2002). Layer-by-layer nanostructured hybrid films of polyaniline and vanadium oxide *J. Nanosci. Nanotech.,* 2(1), 29-32

[231] Pang ,S., Li ,G., Zhang ,Z. (2005). Synthesis of Polyaniline-Vanadium Oxide Nanocomposite Nanosheets, *Macromol. Rapid. Commun.*26(15), 1262-1265

[232] Li, Z.F., Ruckenstein, E. (2002). Intercalation of Conductive Polyaniline in the Mesostructured V_2O_5 Langmuir,18(18), 6956-6961

[233] Karatchevtseva, I., Zhang, Z., Hanna, J., Luca, V. (2006). Electrosynthesis of Macroporous Polyaniline−V_2O_5 Nanocomposites, *Chem. Mater* .18(20), 4908-4916

[234] Malta,M., Louarn,G., Errien, N., Toresi, R.M .(2003). *Electrochem, Commun.*, 5(12), 1011-1015

[235] Malta,M., Louarn, G., Errien ,N., Toresi, R.M. (2006) Redox *J. Power. Sources*,156(2), 533-540

[236] Spahr,M.E., Bitterli, P., Nesper, R., Müller, M., Krumeich, F., Nissen ,H.U. (1998). Redox-Active Nanotubes of Vanadium Oxide, *Angew. Chem. Int. Ed.*, 37(9),1263-1265

[237] Niederberger, M., Muhr, H.J., Krumeich, F., Bieri ,F., Günther, D., Nesper, R. (2000). Low-Cost Synthesis of Vanadium Oxide Nanotubes via Two Novel Non-Alkoxide Routes, *Chem. Mater.,*12(7), 1995-2000

[238] Reisono, J.M., Muhr, H.J., Bieri, F., Nesper,R. (2000).Controlled Uptake and Release of Metal Cations by Vanadium Oxide Nanotubes, *Helv. Chim. Acta.*, 83(8), 1724-1733

[239] Cuentas-Gallegos, A.K., Gómez-Romero, P. (2006) Triple hybrid *J. Power. Sources*, 161(1),580-586

[240] Posudievsky , O.Y., Biskulova, S.A., Pokhodenko, V.D. (2004). New hybrid, *J. Mater. Chem.* 14, 1419-1423

[241] Anaissi, F.J., Demets, G.J.F., Timm, R.A., Toma, H.E. (2003). Hybrid polyaniline, *Mater. Sci. Eng.* A 347(102),374-381

[242] Anaissi, F.J., Demets, G.J-F., Toma, H.E., Vieira, A.C. (1999) Modified electrodes, *J. Electroanal. Chem.*, 464(1),48-53

[243] Posudievsky, O.Y., Kurys, Y.I., Pokhodenko, V.D. (2004). 12-Phosphormolibdic acid, *Synth. Met.*144(2) ,107-111

[244] Rajesh, B., Ravindranathan ,T.K., Bonard, J-M., Xanthopolous, N., Mathieu, H.J., Viswanathan, B. (2002). Pt Supported on Polyaniline-V_2O_5 Nanocomposite as the Electrode Material for Methanol Oxidation, *Electrochem.Solid. State Lett.* 5(12), E71-E74

[245] Rajesh,B., Ravindranathan, T.K., Bonard ,J-M., Mathieu, H.J., Xanthopolous, N., Viswanathan, B. (2005) Electronically conducting hybrid material as high performance catalyst support for electro catalytic, *J. Power Sources,* 141(1), 35-38

[246] SU,S-J., Kuramoto,N. (2000). Processable polyaniline, *Synth. Met.* 114(2), 147-153

[247] Li, X., Chen, W., Bian, C., He, J., Xu, N., Xue, G. (2003). Surface modification of TiO_2, *Appl. Surf. Sci.* 217(1-4), 16-22

[248] Li,X., Wang, G., Li, X., Lu, D .(2004).Surface properties of polyaniline *Appl. Surf. Sci.* 229(104), 395-401

[249] Xu ,J-C., Liu ,W-M., Li, H-L.(2005).Titanium, *Mater Sci. Eng.*C 25 (4), 444-447

[250] Zhang, L., Liu, P., Su ,Z. (2006). Preparation *Polym. Degr. Stab.* 91(9), 2213-2219

[251] Sairam,M., Patil, M.B., Veerapur, R.S., Patil, S.A., Aminabhavi,T.M. (2006). Novel dense poly (vinyl alcohol *J. Membr. Sci.* 281(1-2) , 95-102

[252] Ma,X., Wang,M., Li, G., Chen ,H., Bai, R. (2006), Preparation, *Mater. Chem .Phys.* 98(2-3), 241-247

[253] Somani,P.R., Marimuthu, R., Mulik, U.P., Sainkar,S.R., Amalnerkar, D.P. (1999). High piezoresistivity and its origin in conducting polyaniline, *Synth. Met.*106(1), 45-52

[254] Sui,X., Chu, Y., Xing, S., Yu ,M., Liu, C. (2004). Self, *Colloids. Surf.* A., 251(1-3), 103-107

[255] Sui,X., Chu,Y., Xing ,S., Liu ,C. (2004). Synthesis of PANI, *Mater. Lett.,* 58(7-8), 1255-1259

[256] Dey ,A., De, A., De, S.K. (2004). Characterization and dielectric properties of polyaniline–TiO_2 nanocomposites, Nanotechnology 15(9),1277-1283

[257] Xiong, S., Wang, Q., Xia ,H. (2004). Template synthesis, *Synth. Met.*146(1), 37-42

[258] Lu,X., Yu ,Y., Chen, L., Mao ,H., Gao, H., Wang ,J., Zhang ,W., Wie ,Y. (2005) .Aniline dimer–COOH assisted preparation of well-dispersed polyaniline–Fe_3O_4 nanoparticles, *Nanotechnology* 16(9), 1660-1665

[259] Aphesteguy, J.C., Jacobo, S.E. (2004). Composite of polyaniline, *Physica.* B., 354(1 -4):224-227

[260] Deng, J., He, C., Peng, Y., Wang, J., Long, X., Lei, P., Chan, A.S.C. (2003). Magnetic and conductive, *Synth. Met.*139(2) :295-301

[261] Long, Y., Chen, Z., Duvail, J.L., Zhang, Z., Wan, M. (2005). Electrical and magnetic *Physica.* B., 370(1-4) , 121-130

[262] Xue, W., Fang, K., Qiu, H., Li, J., Mao, W. (2006). Magnetic and conductive, *Synth. Met.,* 156, 506-508

[263] Bao, L., Jiang, J.S. (2005). Evolution of microstructure, *Physica.* B 367(1-4), 182-187

[264] Lu, X., Mao ,H., Chao, D., Zhang, W., Wei, Y. (2006). Ultrasonic synthesis *J. Solid. State Chem.*, 179(8) , 2609-2615

[265] Tang ,B.Z., Geng, Y., Lam ,J.W.L., Li, B. (1999). Processible Nanostructured Materials with Electrical Conductivity and Magnetic Susceptibility: Preparation and Properties of Maghemite/Polyaniline Nanocomposite Films *Chem. Mater.*, 11(6), 1581-1589

[266] Yang, Q.L., Zhai ,J., Song, Y.L., Wan, M.X., Jiang ,L., Xu ,W.G., Li ,Q.S. (2003). Synthesis and Characterization of Conducting Polyaniline/γ-Fe_2O_3 Magnetic Nanocomposite, *Synth. Met.* 135-136, 819-820

[267] Li,G., Yan, S., Zhou ,E., Chen, Y. (2006). Preparation *Colloids Surf.* A .276(1-3), 40-44

[268] Kazantseva, N.E., Vilčáková, J., Křesálek, V., Sáha, P., Sapurina, I., Stejskal, J. (2004) . Magnetic behavior, *J. Magn. Magn. Mater.*, 269(1), 30-37

[269] Maity, A., Biswas, M. (2004). Water-dispersible conducting nanocomposites of binary polymer systems. I. Poly (N-vinylcarbazole)-polyaniline-Al_2O_3 nanocomposite system, *J. Appl. Polym. Sci.* 94(15):803-811

[270] Geng ,L., Zhao ,Y., Huang ,X., Wang, S., Zhang, S., Wu ,S. (2007), Characterization and gas , *Sens. Actuators.* B ,120(2), 568-572

[271] He, Y. (2004).Preparation, *Powder. Technol.,* 147(1-3),59-63

[272] He, Y. (2005). A novel emulsion route to sub-micrometer, *Appl. Surf. Sci* .249(1-4),1-6

[273] He, Y. (2005).Synthesis of polyaniline, *Mater .Chem .Phys* .92(1), 134-137

[274] Gemeay, A.H., Mansour, I.A., El-Sharkawy, R.G., Zaki, A.B. (2005). Preparation, *Eur. Polym .J.* 41(11) :2575-2583

[275] De Castro, E.G., Zarbin ,A.J.G., De Oliveira ,H.P., Galembeck, A. (2004). Novel flexible, freestanding and transparent, *Synth .Met.* 146(1) , 57-62

[276] Khan, A.A., Inamuddin. (2006). Applications of Hg(II) sensitive polyaniline, *Sens. Actuators.* B .120(1) , 10-18

[277] Chang, T.C., Shen, W.Y., Ho, S.Y. (1995). Oxidative polymerization, *Microporous. Mater.* 4(5), 335-343

[278] Liu, Y.J., Kanatzidis, M.G. (1995). Post intercalative Polymerization of Aniline and Its Derivatives in Layered Metal Phosphates, *Chem. Mater.* 7(8), 1525-1533

[279] Wang, J., Hu, Y., Tang ,Y., Chen ,Z. (2003), Preparation, *Mater. Res. Bull.,* 38(8), 1301-1308

[280] Wang ,J., Hu ,Y., Song, L., Chen ,Z. (2005). Sonochemical preparation of nanocomposite of gamma-zirconium, *Solid. State Ionics*, 167(3-4), 425-430

[281] De, S., De ,A., Das, A., De, S.K. (2005). Transport and dielectric properties of α-zirconium, *Mater. Chem. Phys.*, 91(2-3), 477-483

[282] Pethar, S., Patil, R.C., Kher, J.A., Vijayamohanan, K. (1999). Deposition and characterization of CdS nanoparticle/polyaniline, *Thin Solid Films*, 349(1-2) ,105-109

[283] Khanna, P.K., Lonkar, S.P., Subbarao, V.V.V.S., Jun, K-W. (2004). Polyaniline–CdS nanocomposite from organ, *Mater. Chem. Phys.*87 (1): 49-52.

[284] Khanna, P.K., Kulkarni, M.V., Singh, N., Lonkar, S.P. (2006).Synthesis of HCl doped, *Mater. Chem .Phys.*95(1), 24-28

[285] Khiew ,P.S., Huang, N.M., Radiman, S., Soot, A. Md. (2004). Synthesis and characterization of conducting polyaniline, *Mater. Lett.,* 58(3-4) :516-521

[286] Godovsky, D.Y., Varfolomeev, A.E., Zaretsky, D.F., Nayana, C.R.L., Kündig, A., Weder, C., Caseri, W. (2001). Preparation*J. Mater. Chem* .11, 2465-2469

[287] Chandrakanthi, R.L.N., Careem, M.A. (2002).Preparation *Thin Solid Films*, 417(1-2), 51-56

[288] Qiu, Y., Gao, L. (2005). Novel Polyaniline/Titanium *J. Phys. Chem.* B.109(42),19732-19740

[289] Bein, T., Enzel, P. (1989). Encapsulation of polyaniline *Synth. Met.* 29(1), 163-168

[290] Enzel ,P., Bein, T. (1989). Inclusion of polyaniline, *J. Phys. Chem.* ., 93(17), 6270-6272

[291] Frisch, H.L., Song, H., Ma ,J., Rafailovich ,M., Zhu ,S., Yang ,N-L., Yan ,X. (2001). , Antiferromagnetic, *J. Phys. Chem.* B .105(47), 11901-11905

[292] Wu, C-J., Bein, T. (1994). Polyaniline Wires in Oxidant-Containing Mesoporous Channel Hosts *Chem. Mater.,* 6(8),1109-1112

[293] Choi, H.J., Cho, M.S., Ahn ,W.S. (2003) Synthesis of polyaniline, *Synth. Met.*, 135-136, 711-712

[294] Cho, M.S., Choi, H.J., Ahn, W.S. (2004). Enhanced Electrorheology of Conducting Polyaniline Confined in MCM-41 Channels, Langmuir ,20(1), 202-207

[295] Anunziata, O.A., Gómez-Costa, M.B., Sánchez, R.D. (2005).Preparation, *J. Colloid. Interf. Sci.*, 292(2), 509-516

[296] Densakulprasert, N., Wannatong, L., Chotpattananont, D., Hiamtup, P., Sirivat, A., Schwank, J. (2005).Electrical conductivity, *Mat. Sci. Eng.* B.,117(3), 276-282

[297] Cho, M.S., Choi, H.J., Kim, K.K., Ahn, W-S. (2002). Synthesis and Characterization of Polyaniline/Mesoporous SBA-15 Nanocomposite, *Macromol.Rapid. Commun.,* 23(12),713-716

[298] Szleifer, I., Yerushalmi-Rozen, R. (2005). Polymers and carbon, *Polymer*, 46(19), 7803-7818

[299] Coleman, J.N., Curran, S., Dalton, A.B., Davey, A.P., McCarthy, B., Blau, W., Barklie, R.C. (1999). Physical Doping of a Conjugated Polymer with Carbon, *Synth. Met.* 102(1-3) , 1174-1175

[300] Cochet, M., Maser, W.K., Benito, A.M., Callejas, M.A., Martínez, M.T., Benoit, J-M., Schreiber,J., Chauvet ,O. (2001) . Synthesis of a new polyaniline*Chem. Commun.* (16), 1450-1451

[301] Maser ,W.K., Benito, A.M., Callejas, M.A., Seeger ,T., Martínez ,M.T., Schreiber ,J., Muszynski ,J., Chauvet, O., Osváth ,Z., Koós, A.A., Biró, L.P

.(2003) Synthesis and characterization of new polyaniline/nanotube composites, *Mat. Sci. Eng.* C., 23(1-2) :87-91

[302] Deng ,J.G., Ding ,X.B., Peng ,Y.X., Chan, A.S.C. (2001). Hybrid composite of polyaniline containing carbon nanotube Chinese. *Chem. Lett.* 12(12), 1037-1040

[303] Deng, .J, Ding, X., Zhang, W., Peng, Y., Wang, J., Long, X., Li, P., Chan, A.S.C. (2002) ,Carbon *Eur. Polym. J.* 38(12), 2497-2501

[304] Zengin ,H., Zhou, W., Jin ,J., Czerw, R., Smith, D.W.Jr, Echegoyen ,L., Carroll ,D.L., Foulger, S.H., Ballato, J. (2002) .Carbon Nanotube Doped Polyaniline, *Adv. Mater.,* 14(20), 1480-1483

[305] Li ,X-H., Wu, B., Huang ,Ji-E., Zhang, J., Liu, Z-F., Li ,H-L. (2003). Fabrication and characterization of well-dispersed single-walled carbon, *Carbon*, 41(8),1670-1673

[306] Huang ,J-E., Li ,X-H., Xu ,J-C., Li ,H-L. (2003). Well-dispersed single-walled carbon, *Carbon* 41(14), 2731-2736

[307] Guo, D-J., Li, H-L. (2005). Well-dispersed multi-walled carbon, *J .Solid. State Electrochem.* 9(6), 445

[308] Long ,Y., Chen, Z., Zhang ,X., Zhang, J., Liu, Z. (2004). Synthesis and electrical properties of carbon nanotube polyaniline composites, *Appl. Phys. Lett.*, 85, 1796-1798

[309] Zhang ,X., Zhang, J., Wang ,R., Liu ,Z .(2004). Cationic surfactant, *Carbon*,42(8-9) ,1455-1461

[310] King, R.C.Y., Roussel, .F (2005).Morphological and electrical characteristics of polyaniline, *Synth. Met.,* 153(1-3) ,337-340

[311] Yu ,Y., Che ,B., Si, Z., Li ,L., Chen, W., Xue ,G .(2005) .Carbon, *Synth. Met.* ,150(3), 271-277

[312] Wei ,Z., Wan, M., Lin, T., Dai, L. (2003). Polyaniline Nanotubes Doped with Sulfonated Carbon Nanotubes Made Via a Self-Assembly Process, *Adv. Mater.* 15(2), 136-139

[313] Zhang, X., Zhang, J., Liu, Z. (2005). Tubular composite of doped *Appl. Phys*. A 80(8), 1813-1817

[314] Wu, T-M., Lin, Y-W., Wu, C-S., Liao, C-S. (2005). Preparation *Carbon*, 43(4),734-740

[315] Philp, B., Xie ,J., Abraham, J.K., Varadan, V.K. (2004). A new synthetic route to enhance polyaniline assembly on carbon nanotubes in tubular composites, *Smart. Mater. Struct.,* 13(6), N105-N108

[316] Philp, B., Xie, J., Abraham, J.K., Varadan, V.K. (2005). Polyaniline / carbon, *Polym. Bull.*, 53(2),127

[317] Baibarac ,M., Baltog, I., Lefrant, S., Mevellec ,J.Y., Chauvet, O. (2003) , Polyaniline and Carbon, *Chem. Mater.,* 15(21), 4149-4156

[318] Sainz, R., Benito, A.M., Martínez, M.T., Galindo, J.F., Sotres, J., Baró, A.M., Corraze, B., Chauvet, O., Dalton, A.B., Baughman, R.H., Maser, W.K. (2005). A soluble and highly functional polyaniline–carbon nanotube composite *Nanotechnology* 16(5) :S150-S154

[319] Park, H., Zhao, J., Jian, P., Lu, J.P. (2006) Effects of Sidewall Functionalization on Conducting Properties of Single Wall Carbon *Nano. Lett.* 6(5),916-919

[320] Giusca, C., Baibarac, M., Lefrant, S., Chauvet, O., Baltog, I., Manaila, R., Devenyi, A. (2002). C_{60}–polymer, *Carbon* ,40(9),1565-1574

[321] Du, X.S., Xiao, M., Mang, Y.Z. (2004). Facile synthesis *Eur. Polym. J.* 40(7),1489-1493

[322] Du ,X.S., Xiao, M., Mang ,Y.Z. (2004).Synthesis and characterization of polyaniline/graphite conducting nanocomposites, *J. Polym. Sci.* B 42(10), 1972-1978

[323] Bourdo, S. E., Viswanathan ,T. (2005) Graphite/Polyaniline (GP) composites, *Carbon*, 43(14),2983-2988

[324] Luo, K., Guo, X., Shi, N., Sun, C. (2005). Synthesis and characterization of core–shell nanocomposites, *Synth. Met.*,151(3) ,293-296

[325] Theron, A., Zussman, E., Yarin, A.L. (2001). Electrostatic field-assisted alignment of electrospun nanofibres, *Nanotechnology*, 12, 384–90.

[326] Jin, H.J., Fridrikh, S., Rutledge, G.C., Kaplan, D. (2002). Electrospinning Bombyx mori silk with poly (ethylene oxide). Abstracts of Papers American Chemical Society, 224(1–2):408.

[327] Yoshimoto, S., Ohashi ,F., Kameyama T. (2004) Simple Preparation of Sulfate Anion-Doped Polyaniline-Clay Nanocomposites by an Environmentally Friendly Mechano-chemical Synthesis Route, *Macromol. Rapid. Commun.,*25(19), 1687 – 1691

[328] Huang, J., Moore, J.A., Acquaye, J.H. and Kaner R.B. (2005). Mechanochemical Route to the Conducting Polymer Polyaniline, *Macromolecules*, 38 (2), 317–321

[329] Lau, K. K. S.,Bico, J., Teo, K.B.K., Chhowalla, M., Amaratunga,G. A. J., Milne, W. I., McKinley, G. H., Gleason, K. K. (2003). Super hydrophobic *Nano. Lett.,* 3(12), 1701-1705.

[330] Gupta, M., Kapur, V., Pinkerton, N. M., Gleason, K.K. (2008) . Initiated Chemical Vapor Deposition (iCVD) of Conformal Polymeric Nano coatings *Chem. Mater.* 20(4), 1646-1651.

[331] Nishida, H., Yamashita, M., Andou, Y., Jeong, J.-M., Endo, T. (2005). Gas-Phase Assisted Surface Polymerization Behavior of β-Propiolactone on Inorganic and Organic Substrates and Consequent Composite Production, *Macromol. Mater. Eng.*, 290(8), 848-856.

[332] Wang, K.H., Chung, I.J., Jang, M.C., Keum, J.K. Song, H.H.(2002). Deformation Behavior of Polyethylene/Silicate Nanocomposites Macromolecules 35 (14), 5529–5535.

[333] Akita, H., Hattori T. (1999). Studies on molecular composite. I. Processing of molecular composites using a precursor polymer for poly(p-phenylene benzobisthiazole) *J. Polym. Sci. B: Polym. Phys*. 37(3), 189–197.

[334] Akita, H., Kobayashi, H. (1999).Studies on molecular composite. III. Nano composites consisting of poly (p-phenylene benzobisthiazole) and thermoplastic polyamide *J. Polym. Sci. B: Polym. Phys*. 37 (3) 209–218.

[335] Akita, H., Kobayashi, H., Hattori, T., Kagawa, K., (1999) .Studies on molecular composite. II. Processing of molecular composites using copolymers consisting of a precursor of poly(p-phenylene benzobisthiazole) and aromatic polyamide, *J. Polym. Sci. B: Polym.Phys*., 37(3), 199–207.

[336] Chang, J.-H. An, Y.U. (2002). Nanocomposites of polyurethane with various organo clays: Thermo mechanical properties, morphology, and gas permeability *J. Polym. Sci. B: Polym. Phys*. 40 (7), 670–677.

[337] Zavyalov, S.A., Pivkina, A.N., Schoonman, J. (2002). Formation and characterization of metal-polymer, *Solid State Ionics*, 147(3-4) , 415–7419.

[338] Park, J.H., Jana, S.C. (2003) Mechanism of Exfoliation of Nanoclay Particles in Epoxy - Clay Nanocomposites, *Macromolecules*, 36(8) 2758–2768.

[339] Koo, C.M. , Kim, M.J. , Min, H., Kim, S.O. , Chung, I.J. (2003).Mechanical and rheological properties of the maleated polypropylene-layered silicate nanocomposites with different morphology, *J. Appl. Polym. Sci.,* 88 (6), 1526–1535.

[340] Jordan, J. Jacob, K.I., , Tannenbaum, R.,. Sharaf, M.A., and Jasiuk I.(2005).Experimental trends in polymer nanocomposites—a review, Materials Science and Engineering A,393(1-2),1-11

[341] Ranade, A., D'Souza, N.A., Gnade, B. (2002). Exfoliated and intercalated polyamide, *Polymer,* 43(13), 3759–3766.

[342] Mativetsky, J.M., Datars, W.R.(2002).Morphology and electrical properties, *Physica* B, 324(1-4), 191-204

[343] Ballarin, B., Facchini, M., Dal Pozzo, L., Martini, C. (2003). Comparison of different porous *Electrochem. Commun.* 5(8) 625 -631

[344] Committee on Technology, National Science and Technology Council: "National Nanotechnology Initiative", A Report by the Interagency Group on Nanoscience, Engineering and Technology (Washington, D.C., 2000).

[345] Bockrath, M., Cobden, D. H., McEuen, P. L., Chopra, N. G., Zettl, A., Thess, A., Smalley, R. E. (1997), Single-Electron Transport in Ropes of Carbon Nanotubes. *Science*, 275, 1922-1999

[346] Pinto, N. J., Johnson, A. T. Jr., MacDiamid, A.G., Mueller C.H., Theofylaktos, Robinson, N., D. C. and Miranda, F.A. (2003). Electrospun polyaniline/ polyethylene oxide nanofiber field-effect transistor *Appl. Phys. Lett.* 83(20) 4244-4246

[347] Sze, S. M. (1985). Semiconductor Devices Physics and Technology (Wiley, New York).

[348] Horowitz, G. (1998). Organic Field-Effect Transistors *Adv. Mater*. 10(5), 365-377.

[349] Brown, A. R. , Jarrett, C. P., de Leeuw, D. M. and Matters, M. (1997). Field-effect transistors *Synth. Met*. 88(1) , 37-55

[350] Bezryadin, A., Verschueren, A. R. M., Tans, S. J. and Dekker, C. (1998) Multiprobe transport experiments on individual single-wall carbon nanotubes. *Phys. Rev. Lett.* 80, 4036–4039.

[351] Keren, K., Berman, R. S., Buchstab, E., Sivan, V. And Braun, E. (2003). DNA-Templated Carbon Nanotube Field-Effect Transistor Science, 302, 1380-1382.

[352] Quali, L., Krasnik V.V., Stalmach, U. and Hadziioannou, G.,(1999). Oligo (phenylenevinylene)/Fullerene Photovoltaic Cells: Influence of Morphology *Adv. Mater*., 11(18) , 1515-1518

[353] Peters, E., Avan P.H., Knol, J., Brabec, C.J., Saricftci, N.S., Hummelen, J.C. and Janssen, R.A.J. (2000). Synthesis, Photo physical Properties, and Photovoltaic Devices of Oligo(p-phenylene vinylene)-fullerene*J. Phys. Chem.* B 104(44) , 10174-10190

[354] Curran, S.A, . Ajayan, P.M., Blau, W.J., Carroll, D.L., Coleman, J.N., Dalton, A.B., Davey, A.P., Drury , A., McCarthy , B., Maier , S., and Strevens, A.(1998). A Composite from Poly (m-phenylenevinylene-co-2,5-dioctoxy-p-phenylenevinylene) and Carbon Nanotubes: A Novel Material for Molecular Optoelectronics *Adv. Mater*. 10(14), 1091-.1093

[355] Romero, D.B. Carrard, M. Heer, W.de and Suppiroli, L. (1996).A carbon nanotube/organic semiconducting polymer heterojunction *Adv. Mater*. 8, 899-903

[356] Kymakis, E. and Amaratunga, G.A.J. (2002). Single-wall carbon nanotube/conjugated polymer photovoltaic devices, *Appl. Phys. Lett.* 80(1), 112-114

[357] Matsumura, M., Ito, A., and Miyamae, Y. (1999).Accumulation of positive charges in organic light-emitting diodes with a double-layer structure *Appl. Phys. Lett.* 75(8) ,1042-1044

[358] Brown, T.M. Kim, J.S., Friend, R.H., Cacialli, F., Daik, R. and Feast, W.J.(1999).Built-in field electro absorption spectroscopy of polymer light-emitting diodes incorporating a doped poly(3,4-ethylene dioxythiophene) hole injection layer *Appl. Phys. Lett.* 75(12), 1679-1681.

[359] Woo, H.S., Czerw, R. ,Webster, S., Carroll, D.L., Ballato, J., Strevens, A.E., Brien, D.O. and Blau, W.J.(2000).Hole blocking in carbon nanotube–polymer composite organic light-emitting diodes based on poly (m-phenylene vinylene-co-2, 5-dioctoxy-p-phenylene vinylene), *Appl. Phys. Lett.*, 77(9), 1393-1395.

[360] Woo, S., Czerw, R., Webster, S., Carroll, D.L., Park J.W., and Lee, J.H.(2001) *Synth. Met.*, 116(1-3), 369-372

[361] Kim, J-Y. , Kim, M., Kim, H.M., Joom, J.and Choi J.H. (2002). Electrical and optical *Opt. Mater.* 21(1-3) , 147-151

[362] Granstrom, M., Petritsch, K., Arias, A.C., Lux, A., Anderson, M.R. and Friend, R.H. (1998). Laminated fabrication of polymeric photovoltaic diodes *Nature* 395(6699), 257-260

[363] Camaioni, N., Ridolfi, G., Casalbore-Miceli, G., Possamai, G., Garlaschelli, L.and Maggini, M. (2003). A stabilization, *Sol. Energ. Mat. Sol.* C ., 76(1), 107-113

[364] Halls, J.J.M. and Friend,R.H. (1997) The photovoltaic effect in a poly(p-phenylenevinylene)/perylene heterojunction,*Synth. Met.*,85(1-3),1307-1308

[365] Schadler, L.S., Giannaris, S.C.and Ajayan, P.M.(1998). Load transfer in carbon nanotube epoxy composites *Appl. Phys. Lett.*73 (26), 3842-3844.

[366] Salvetat, J. Briggs, G.A.D., Bonard, J., Bacsa, R.R., Kulik, A.J., Stockli, T., Burnham, N.A.and Forro, L.(1999) Elastic and Shear Moduli of Single-Walled Carbon *Phys. Rev. Lett.*, 82(5), 944-947 .

[367] Berber, S.Kwon, Y., and Tomanek, D. (2000). Unusually High Thermal Conductivity, *Phys. Rev.Lett.,* 84(2) , 4613-4616

[368] Baibarac, M. and Gómez-Romero, P.(2006) Nanocomposites Based on Conducting Polymers and Carbon Nanotubes from Fancy Materials to Functional Applications, *J. Nano. Sci. Nanotech.*, 6, 1–14

[369] Kong, J., Franklin, N.R., Zhou, C., Chapline, M.G., Peng, S., Cho, K. and Dai, H. (2000). Nanotube Molecular Wires as Chemical Sensors Science 287, 622-625

[370] Collins, P.G., Bradley, K., Ishigami, M. and Zettl, A. (2000). Extreme Oxygen Sensitivity of Electronic Properties of Carbon Nanotubes Science 287,1801-1804

[371] Qi, P., Vermesh, O., Grecu, M., Javey, A., Wang, Q. and Dai, H. (2003). Toward Large Arrays of Multiplex Functionalized Carbon, *Nano. Lett.*, 3(3), 347-351

[372] An, K.H., Jeong, S.Y., Hwang, H.R .and Lee, Y.H. (2004). Enhanced Sensitivity of a Gas Sensor Incorporating Single-Walled Carbon Nanotube-Polypyrrole Nanocomposites, *Adv. Mater.*16(12) ,1005-1009

[373] Valentini, L., Bavastrello, V., Stura, E., Armentano, I., Nicolini, C., and Kenny, J.M.(2004). Sensors for inorganic, *Chem. Phys. Lett.*, 383(5-6), 617-622.

[374] Bavastrello, V., Stura, E., Carrara, S., Eroklin, V. and Nicolini, C. (2004). *Sensors Actuat. B: Chem.*, 98(2-3), 247-253.

[375] Sotiropoulou, S., Gavalas, V., Vamvakaki, V. and Chaniotakis, N.A. (2003). Novel carbon, *Biosens. Bioelectron.* 18(2-3) , 211-215

[376] Wang, S.G., Zhang, Q., Wang, R., Yoon, S.F., Ahn, J., Yang, D.J., Tian, J.Z., Li, J.Q., and Zhou, Q. (2003).Multi-walled carbon, *Electrochem. Commun.* 5(9), 800-803

[377] Gao, M., Dai, L. and Wallace, G.G. (2003). Biosensors Based on Aligned Carbon Nanotubes Coated with Inherently Conducting Polymers Electroanal., 15(13) , 1089-1094

[378] Cai, H., Xu, Y., He, P.G. and Fang, Y.Z. (2003). Indicator Free DNA Hybridization Detection by Impedance Measurement Based on the DNA-Doped Conducting Polymer Film Formed on the Carbon Nanotube Modified Electrode Electroanal. 15(23-24) , 1864-1870

[379] Carrara, S., Bavastrello, V., Ricci, D., Sutra, E. and Nicolini, C. (2005). Methods to fabricate, Sensors. *Actuat. B:Chem.* 105(2), 542-548

[380] Cho, S.J., Song, K.S., Kim, J.W., Kim, T.H. (2002). Choo, K. Fuel Chemistry Division, 224th National Meeting of the American Chemical Society, 47, 790.

[381] Eric D.K.(1992) Nanosystems: molecular machinery, manufacturing, and computation, New York: John Wiley and Sons

[382] Maynard, A. (2007).Nanotechnology: The Next Big Thing, or Much Ado about Nothing, *Ann. Occu. Hyg.* 51(1), 1-12

[383] Webster P. (2005) World nanobiotechnology market. *Frost and Sullivan*

[384] Bavastrello, V., Carrara, S., Ram, M.K. and Nicolini, C.(2004). Optical and Electrochemical Properties of Poly(o-toluidine) Multiwalled Carbon, Langmuir, 20(3), 969-973

[385] Li, H.X., Wu, B., Huang, J.E., Zhang, J., Liu, Z.F.and Li, H.L.(2002). Fabrication and characterization of well-dispersed single-walled carbon, *Carbon* 41(8), 1670-1673.

[386] Hirsch, A. (2002).Functionalization of Single-Walled Carbon Nanotubes Angew. *Chem. Ind. Ed.* 41(11), 1853-1859

[387] Lefrant, S., Baltog, I., Baibarac, M., Mevellec, J.Y .and Chauvet, O. (2002). SERS studies on single-walled carbon, *Carbon* 40(12), 2201 -2211

[388] Quillard, S., Louarn, G., Lefrant, S. and MacDiarmind, A.G.,(1994). Vibrational analysis of polyaniline: A comparative study of leucoemeraldine, emeraldine, and pernigraniline bases, *Phys. Rev.* B 50, 12496-12508

[389] Corio, P., Santos, P.S., Brar , V.W ., Samsonidze, G.G., Chou, S.G. and Dresselhaus, M.S. (2003). Potential dependent surface Raman, *Chem. Phys. Lett.* 370(5-6) , 675-682

[390] McCarthy , B., Coleman, J.N., Czerw , R., Dalton, A.B., Carroll, D.L.and Blau, W.J. (2001). Microscopy studies of nanotube *Synth. Met.* 121(1-3) , 1225 -1226

[391] Fitzsimmons, R.J., Strong, D.D., Mohan, S., Baylink, D. J. (1992).Low-amplitude, low-frequency electric field-stimulated bone cell proliferation may in part be mediated by increased IGF-II release, *J. of cellular Physiology*, 150(1) , 84-89

[392] Devoret, M. H., Esteve, D., Urbina, C. (1992). Single-electron transfer in metallic nanostructures *Nature*, 360(6404), 547-552.

[393] Ozin, G. A. (1992).Nanochemistry: Synthesis in diminishing dimensions *Adv. Mater.*, 4(10), 612-619.

[394] Gref, R., Minamitake, Y., Peracchia, M.T., Trubetskoy, V., Torchilin, V., Langer, R. (1994). Biodegradable long-circulating polymeric nano spheres *Science*, 263, 1600-1603.

[395] Parthasarathy, R.V., Martin, C. R. (1994). Synthesis of polymeric microcapsule arrays and their use for enzyme immobilization, *Nature,* 369(6478), 298-300.

[396] Whitesides, G. M., Mathias, J. P., Seto, C. T. (1991). Molecular self-assembly and nano chemistry: a chemical strategy for the synthesis of nanostructures, *Science,* 254, 1312-1319

[397] Schmidt, C. E. , Shastri , V. R., Vacanti, J. P. , Langer, R. (1997) Stimulation of neurite outgrowth using an electrically conducting polymer, *Proc. Natl. Acad. Sci. U.S.A.*, 94(17) , 8948-8953

[398] Groenendaal, B. L., Jonas, F., Freitag, D., Pielartzik, H., Reynolds, J.R. (2000). Poly (3,4-ethylenedioxythiophene) and Its Derivatives: Past, Present, and Future *Adv. Mater.*, 12(7), 481-494.

[399] Smela, E. (1999). Micro fabricated Movable Electrochromic "Pixel" Based on Polypyrrole *Adv. Mater.*, 11(16), 1343-1345

[400] Smela, E. (2003).Conjugated Polymer Actuators for Biomedical Applications, Adv.Mater.15 (6), 481-494.

[401] Gandhi, M. R., Murray, P., Spinks, G. M. , Wallace, G.G. (1995). Mechanism of electromechanical actuation *Synth. Met.*, 73(3), 247-256.

[402] Baughman, R. H. , Shacklette, R. L. , Elsenbaumer, R. L. , Plichta, E. J. , Becht, C. (1990). Conjugated Polymeric Materials: Opportunities in Electronics, Optoelectronics, and Molecular Electronics, (Eds: J. L. Bredas,R. R. Chance), Vol. 182, NATO ASI Series E: Applied Science, pp. 559.

[403] Pei, Q. B. , Inganas, O. (1992). Conjugated polymers and the bending cantilever method: Electrical muscles and smart devices *Adv. Mater.*, 4(4), 227-228.

[404] Otero, T. F. , Cortes, M. T. (2003).Artificial Muscles with Tactile Sensitivity *Adv. Mater.*, 15(4), 279-282.

[405] Adeloju, S. B. Shaw, S. J., Wallace, G. G. (1997). Pulsed-amperometric detection *Anal. Chim. Acta*, 341(2-3), 155-160.

[406] Pernaut, J. M. , Reynolds, J. R. ,(2000). Use of Conducting Electro active Polymers for Drug Delivery and Sensing of Bioactive Molecules. A Redox *J. Phys. Chem.* B., 104(17) , 4080-4090

[407] Reinhard, C. S. , Radomsky, M. L. , Saltzman, W. M. , Hilton, J., Brem, H. (1991) Polymeric controlled release of dexamethasone *J. Cont. Release*, 16(3), 331-339.

[408] Kim, D. H. , Abidian, M. , Martin, D. C. (2004) .Conducting polymers grown in hydrogel scaffolds coated on neural prosthetic devices, *J. Biomed. Mater. Res.*, 71A (4), 577-585.

[409] Yang, J.Y., Martin, D.C. (2004). Microporous conducting polymers, *Sens. Actuators*, B, 101(1-2), 133-142.

[410] Kipke, D. R. , Vetter, R. J. , Williams, J. C. , Hetke, J. F. , (2003) .Silicon-substrate intracortical microelectrode arrays for long-term recording of neuronal spike activity in cerebral cortex, *Neur. Sys. Reh.* 2003, 11(2), 151-155

[411] Zong, X. H., Kim, K., Fang, D. F., Ran, S.F., Hsiao, B.S., Chu, B. (2002). Structure and process relationship, *Polymer*, 43(16), 4403-4412

[412] Ramanaviciene, A., Finkelsteinas,A., Ramanavicius,A. (2004) Molecularly Imprinted Polypyrrole for Sensor Design, Materials Science (Medziagotyra) 10, 18-20

[413] Ramanaviciene, A., Ramanavicius, A., (2002) Application of polypyrrole for the creation of immunosensors, *Crit. Rev. Anal. Chem.*32, 245-252

[414] Ramanaviciene, A., Stalnionis, G., Ramanavicius, A. (2004) Piezoelectric affinity biosensor for diagnosing bovine leukemia. *Biologija1*, 33-35

[415] Hernández, R. M., Richter, L., Semancik, S. and Mallouk, T. E. (2004). Template Fabrication of Protein-Functionalized Gold–Polypyrrole–Gold Segmented Nanowires, *Chem. Mater.*, 16(28) , 3431-3438

[416] Smith, P.A., Nordquist, C. D. , Jackson, T. N. , Mayer, T. S. , Martin , B. R., Mbindyo, J.. and Mallouk, T. E. (2000). Electric-field assisted assembly and alignment of metallic nanowires, *Appl. Phys. Lett.*, 77(9),1399-1401

[417] Crespi, H. (2004). Catalytic Nanomotors: Autonomous Movement of Striped Nanorods, *J. Am. Chem. Soc.*, 126 (41), 13424-13431

[418] Di, J., Sogah, D. Y. (2006). Intergallery Living Polymerization Using Silicate-Anchored Photoiniferter. A Versatile Preparatory Method for Exfoliated Silicate Nanocomposites, *Macromolecules,* 39(3), 1020-1028.

[419] Ericson, Lars, M. (2000). Strength Characterization of Suspended Single-Wall Carbon Nanotube Ropes, Masters Thesis, Rice University, Oct 2000.

[420] Jana, S.C. and Jain, S. (2001). Dispersion of Nanofillers in High Performance Polymers using Reactive Solvents as Processing Aids, *Polymer*, 42, 6897–6905

[421] Zhang, B., Xie, C., Hu, J., Wang, H. and Gui, Y.H. (2006). Novel 1–3 Metal Nanoparticle/Polymer Composites Induced by Hybrid External Fields, Composites Science and Technology, 66, 1558–1563.

[422] Nanoscience and Nanotechnologies, July 2004, the Royal Society and the Royal Academy of Engineering. (Reproduced in part with permission 2005.)

[423] Muller, J., Huaux, F. and Lison, D. (2006). Respiratory Toxicity of Carbon Nanotubes: How Worried Should We Be?, *Carbon,* 44: 1048–1056

[424] Warheit, D.B. (2006). What is Currently Known About the Health Risks Related to Carbon Nanotube Exposures?, *Carbon*, 44: 1064–1069.

[425] Chengang, C. and David, C. (2003). Processing and Morphological Development of Montmorillonite Epoxy Nanocomposites, *Nanotechnology,* 14, 643–648.

[426] Toensmeier, P.A. (2004). Nanotechnology Faces Scrutiny Over Environment and Toxicity, Plastic.Engg., 14-17

[427] Borm, P.J.A. (2002). Particle Toxicology: From Coal to Nanotechnology, *Inhal. Toxicol.*,14, 311– 324.

[428] Donaldson, K., Stone, V., Tran, C.L., Kreyling, W. and Borm, P.J.A. (2004). *Nanotoxicology, Occup. Environ. Med.* 61: 727-728

[429] Agarwal, S., Tatli, E., Clark, N.N. and Gupta, R. (2005). International Symposium on Polymer Nanocomposites Science and Technology, Boucherville, Canada, September, 2005

[430] Koyama, S., Endo, M., Kim, Y.A., Hayashi, T., Yanagisawa, T., Osaka, K., Koyama, H., Haniu, H. and Kuroiwa, N. (2006). Role of Systemic T-cells and Histopathological Aspects after Subcutaneous Implantation of Various Carbon Nanotubes in Mice, *Carbon*, 44(6): 1079–1092.

[431] Shvedova, A.A., Kison, E.R. and Mercer, R. (2005). Unusual Inflammatory and Fibrogenic Pulmonary Responses to Single Walled Carbon Nanotube in Mice, *Am. J. Physiol. Lung Cell. Mol. Physiol.*, 289.

[432] Warheit, D.B. (2004). Nanoparticles: Health Impact?, Materials Today, 7(2), 32–35

[433] Monteiro-Riviere, N.A., Nemanich, R.J., Inman, A.O., Wang, Y.Y. and Riviere, J.E. (2005).Multi-walled Carbon Nanotube Interactions with Human Epidermal Keratinocytes, *Toxicology Letters*, 155, 377–384.

[434] Lademann, J., Weigmann, H.J., et al. (1999). Penetration of Titanium dioxide Micro particles in a Sunscreen Formulation into the Horny Layer and the Follicular Orifice, Skin Pharmacol. *Appl. Skin Physiol.*, 12, 247–256

[435] Dunford, R., Salinaro, A., Cai, L., Serpone, N., Horikoshi, S., Hidaka, H. and Knowland, J. (1997). Chemical Oxidation and DNA Damage Catalyzed by Inorganic Sunscreen Ingredients, *FEBS Lett.* , 418(1–2), 87–90.

[436] Smart, S.K., Cassady, A.I., Lu, G.Q. and Martin, D.J. (2006). The Biocompatibility of Carbon Nanotubes, *Carbon,* 44(6), 1034–1047.

[437] Fiorito, S., Serafino, A., Andreola, F. and Bernier, P. (2006). Effects of Fullerenes and Single wall Carbon Nanotubes on Marine and Human Macrophages, *Carbon,* 44(6), 1100–1105

[438] Hurt, R.H., Monthioux, M. and Kane, A. (2006). Toxicology of Carbon Nanomaterials: Status, Trends, and Perspectives on the Special Issue, *Carbon,* 44(6), 1028–1033.

[439] Donaldson, K., Aitken, R., Tran, L., Stone, V., Duffin, R., Forrest, G. (2006). Carbon nanotubes: a review of their properties in relation to pulmonary toxicology and workplace safety. *Toxicol. Sci.*, 92, 5–22.

[440] Elder, A., Gelein ,R., Silva, V., Feikert, T., Opanashuk, L., Carter, J.(2006). Translocation of inhaled ultrafine manganese oxide particles to the central nervous system. *Environ. Health Perspect.* 114, 1172–1178.

[441] Lam ,C.W., James ,J.T., McCluskey, R., Hunter, R.L. (2004). Pulmonary toxicity of single-wall carbon nanotubes in mice 7 and 90days after intratracheal instillation, *Toxicol. Sci.* 77, 126–134.

[442] Oberdörster, G., Oberdörster, E., Oberdörster, J. (2005).Nanotoxicology: an emerging discipline evolving from studies of ultrafine particles, Environ.Health. *Perspect,* 113, 823–837.

[443] Mark, D. (2004). Nanomaterials—A Risk to Health at Work? Buxton, UK: Health and Safety Laboratory, SK17 9 JN. Available: www.hsl.gov.uk/capabilities/nanosymrep_

[444] Aitken, R.J., Creely, K.S., Tran, C.L. (2004). Nanoparticles: An Occupational Hygiene Review. Health Safety Executive, Research Report 274. London, HSE Books

[445] Heinrich, U., Fuhst ,R., Rittinghauseen ,S., Creutzenberg, O.,Bellmann, B., Koch ,W.(1995). Chronic inhalation exposure of wistar rats and 2 different strains of mice to diesel engine exhaust, carbon black, and titanium dioxide, *Inhal. Toxicol.,* 7, 533–556.

[446] Kipen, H.M., Laskin, D.L. (2005). Smaller is not always better: nanotechnology yields nanotoxicology, *Am. J. Physiol .Lung. Cell.Mol. Physiol.*, 289(5), L696–L697.

[447] Radomski ,A., Jurasz ,P., Alonso-Escolano ,P., Drew, M., Morandi, M.,Tadeusz ,M.(2005). Nanoparticle-induced platelet aggregation and vascular thrombosis. Br J .Pharmacol., 146, 882–893

[448] National Academy of Engineering. (2004). Emerging Technologies and Ethical Issues in Engineering: Papers from a Workshop,14–15 October 2003, http://newton.nap.edu/books/

[449] Wardak ,A., Rejeski, D. (2003). Nanotechnology and Regulation: A Case Study Using the Toxic Substances Control Act (TSCA);A Discussion Paper. Publication 2003–6. Washington, DC, Woodrow Wilson International Center for Scholars, Foresight and Governance Project.

[450] Kantrowitz, A. (1995). The separation of facts and values. Risk 6:105.Available: http:///www.piercelaw.edu/Risk/vol6/spring/Kantro2.htm

[451] Schrader-Frechette, K. (1994). Ethics of Scientific Research, Lanham, MD, Ronan and Littlefield.

[452] Weil, V. (2002). Making sense of scientists' responsibilities at the interface of science and society, *Sci.Eng.Ethics*, 8, 223–227.

[453] Cairns, J Jr. (2003). Integrating top-down/bottom-up sustainability strategies: an ethical challenge. Ethics Sci Environ Politics1–6. http://www.int-res.com/articles/esep/2003/E26

[454] Gert, B., Culver, C.M., Clouser, K.D. (1997). Bioethics: A Return to Fundamentals. New York, Oxford University Press

[455] Samuels, S. (2003). Occupational medicine and its moral discontents, *J. Occup. Environ. Med.*, 45, 1226–1233.

[456] Gallagher, C. (1997). Types of health and safety management systems. In: Health and Safety Management Systems, An Analysis of System Types and Effectiveness. Canberra, ACT, Office of the Australian Safety and Compensation Council.

[457] Jensen, P.L. (2002). Assessing assessment: the Danish experience of worker participation in risk assessment. *Econ. Ind. Democracy* 23, 201–207.

INDEX

A

B

C

D

E

F

G

H

I

J

K

L

M

N

O

P

Q

R

S

T

U

V

W